W9-BJV-088

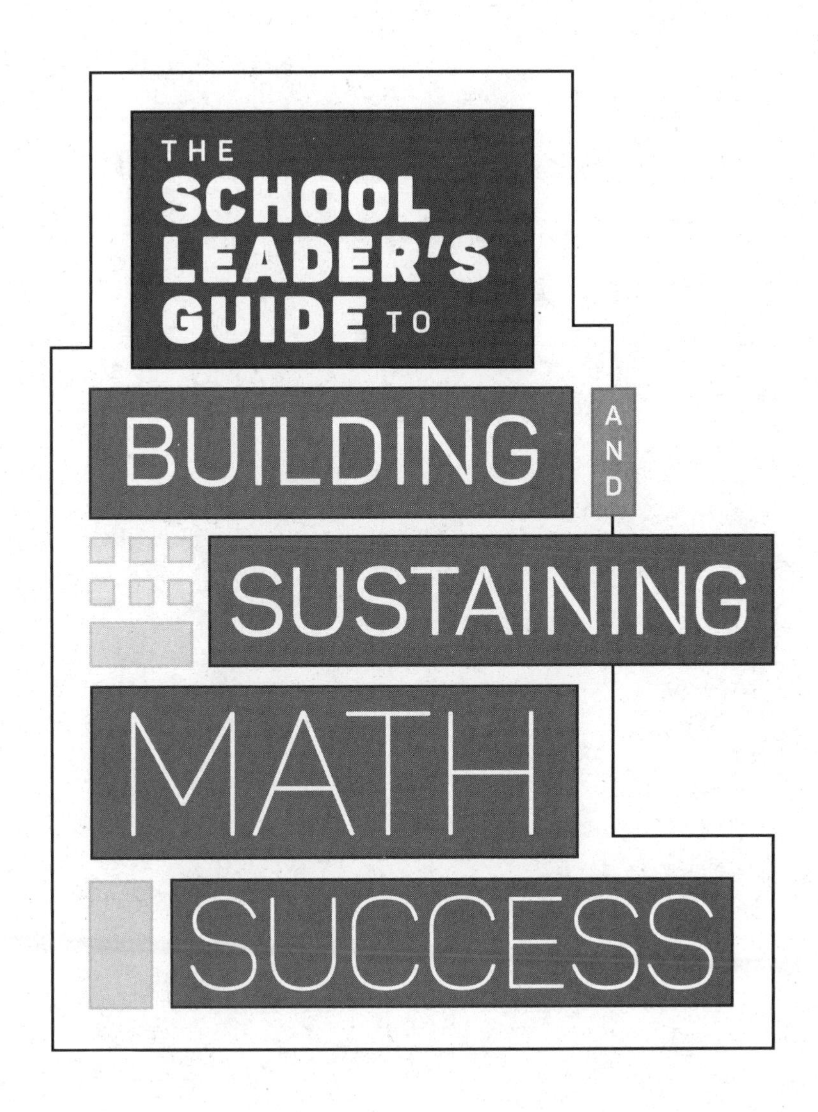
THE
SCHOOL
LEADER'S
GUIDE TO
BUILDING
AND
SUSTAINING
MATH
SUCCESS

MARIAN SMALL
DOUG DUFF

THE SCHOOL LEADER'S GUIDE TO BUILDING AND SUSTAINING MATH SUCCESS

ASCD Alexandria, Virginia USA

1703 N. Beauregard St. • Alexandria, VA 22311-1714 USA
Phone: 800-933-2723 or 703-578-9600 • Fax: 703-575-5400
Website: www.ascd.org • E-mail: member@ascd.org
Author guidelines: www.ascd.org/write

Deborah S. Delisle, *Executive Director;* Stefani Roth, *Publisher;* Genny Ostertag, *Director, Content Acquisitions;* Susan Hills, *Acquisitions Editor;* Julie Houtz, *Director, Book Editing & Production;* Liz Wegner, *Editor;* Judi Connelly, *Associate Art Director;* Melissa Johnston, *Graphic Designer;* Shajuan Martin, *E-Publishing Specialist;* Mike Kalyan, *Director, Production Services;* Absolute Services, *Typesetter;* Sue Curran, *Production Specialist*

PAPERBACK ISBN: 978-1-4166-2638-1 ASCD product #118039 n7/18
PDF E-BOOK ISBN: 978-1-4166-2693-0

Quantity discounts are available: e-mail programteam@ascd.org or call 800-933-2723, ext. 5773, or 703-575-5773. For desk copies, go to www.ascd.org/deskcopy.

Library of Congress Cataloging-in-Publication Data
Names: Small, Marian, author. | Duff, Douglas, author.
Title: The school leader's guide to building and sustaining math success / Marian Small and Douglas Duff.
Description: Alexandria, VA : ASCD, 2018. | Includes bibliographical references and index.
Identifiers: LCCN 2018012380 | ISBN 9781416626381 (pbk.)
Subjects: LCSH: Mathematics—Study and teaching—United States. | Educational leadership—United States. | School environment—United States.
Classification: LCC QA13 .S57 2018 | DDC 510.071—dc23 LC record available at https://lccn.loc.gov/2018012380

27 26 25 24 23 22 21 20 19 18 1 2 3 4 5 6 7 8 9 10 11 12

THE **SCHOOL LEADER'S GUIDE** TO BUILDING AND SUSTAINING MATH SUCCESS

Introduction: Attending to Math Performance 1

1. What Does a Great Math School Look Like? 5

2. Laying the Foundation for a Strong Math Culture in Your School: What Are the Critical Steps? 19

3. Overcoming Resistance and Getting Buy-In 51

4. Looking at Students for Signs of a Strong Math Culture 64

5. Looking at Teachers for Signs of a Strong Math Culture in the Classroom 87

6. Looking at Teachers for Signs of a Strong Math Culture Outside the Classroom 111

7. Monitoring and Sustaining Improvement 131

Appendix A. Look-Fors, Questions, and Feedback for Students 155

Appendix B. Look-Fors, Questions, and Feedback for Teachers in the Classroom 158

Appendix C. Look-Fors, Questions, and Feedback for Teachers Outside the Classroom 160

Further Reading 162

References 163

Index 167

About the Authors 172

Introduction: Attending to Math Performance

There has been increasing international attention to student performance in mathematics, both within the education community and from outside. Within the education sector, there are large numbers of studies each year that focus on the teaching and learning of mathematics. Whether the studies form the basis for doctoral dissertations or whether they are carried out by established educational researchers, the number of studies grows each year, and, with easy access to the internet, the audience for those studies has broadened.

In addition, departments of education, ministries of education, and organizations like the National Governors Association, which supported the development of the Common Core State Standards, are paying increased attention to mathematics teaching and learning. In some jurisdictions, math has become a major focus for entire cities or states.

Even outside of the world of education, the results of high-stakes tests, particularly when used to compare educational jurisdictions, dominate the media in ways that did not occur in the past (e.g., Gordon, 2016; Green, 2014; "PISA Envy," 2013; Stutz, 2013). This has drawn a significant amount of attention to mathematics teaching and learning in the K–12 sector.

While there are vocal critics of how mathematics is currently taught, there are also equally vocal proponents; no matter, the attention is not going away. School district officials and the principals in those districts feel that pressure and the need to improve the performance of their students.

What Changes Performance in a School?

There is abundant evidence from research to support the assertion that schools change when leadership focuses on that change. According to Fullan (2006), effective change occurs when these characteristics are present:

- Motivation
- Capacity building, with a focus on results (building collective knowledge and providing resources to increase the effectiveness of the group in closing the gap in student learning)
  - Learning in context (opportunities to practice in the classroom the ideas being discussed)
  - Changing context (sharing ideas with others in the system, thereby changing the context in which the school operates)
- A bias for reflective action (purposeful thinking about potential changes on the part of the leader and the group)
- Tri-level (broad) engagement (school and community, district, and state or province)
- Both persistence and flexibility in staying the course

However, those who write about how to accomplish this change generally are not grounded in the discipline of mathematics and rarely use examples with mathematics. As a result, principals have difficulty translating change theory into actions to improve math instruction. The change ideas make sense to the leader theoretically, but the implementation is not targeted toward teachers' mathematical practice.

The Purpose of This Resource

Principals who must lead the change to improve student math performance in their schools often tell us that they feel disadvantaged because they are uncomfortable with math themselves. Although they may have successfully completed math courses in school, many remember being taught in ways that are no longer advocated. Therefore, they don't always have much direct experience with math instruction that is based on more recent research, and they don't always see clearly how to lead change that promotes newer approaches to higher-quality math instruction. This resource is meant to provide school and system leaders with a better understanding of what these changes look like and to give specific suggestions for what they can do and what they can ask of their staff and students.

The resource will offer extensive guidance that will be useful to the principal, including advice about the kinds of data to collect and how to use those data to promote teacher change in the teaching of mathematics. In addition, suggestions for "look-fors" that principals can use in their observations and conversations with both students and teachers are provided.

This resource contains significant discussion about the need for a whole-school improvement focus based on student performance. Principals need to instill the belief in all staff that positive change will occur through consistent implementation of adopted core priorities. A collaborative culture of mathematics learning and data literacy must become valued by all staff.

The book also covers the use of sophisticated quantitative data analysis as an ongoing basis for school and class formative assessment. It stresses the importance of monitoring short- and long-term growth, analyzing improvement trends at several points throughout the school year, and tracking individual-level growth from grade to grade. The book describes a culture in which regular staff communication occurs around student performance and whole-school data on multiple platforms.

Included in the resource are examples of many specific scenarios a principal might face when moving a school's math culture, with suggestions for actions to take, suggestions to make, and feedback to give. There are numerous charts that principals might use in staff meetings or with individual teachers in focused discussions. We firmly believe that it is the principal of a school who makes the greatest difference in building real math success in that school. We hope that the advice we offer will help you be that principal.

1 What Does a Great Math School Look Like?

Many of us are familiar with the changes in math instruction that have occurred over the last 20 or 30 years. These have largely been influenced by the National Council of Teachers of Mathematics (NCTM), an organization in the United States and Canada that promotes improved mathematics instruction. Its *Principles and Standards for School Mathematics* (NCTM, 2000), followed later by *Curriculum Focal Points for Prekindergarten Through Grade 8 Mathematics: A Quest for Coherence* (NCTM, 2006), *Focus in High School Mathematics: Reasoning and Sense Making* (NCTM, 2009), and more recently *Principles to Actions: Ensuring Mathematical Success for All* (NCTM, 2014), have served to direct many mathematics initiatives in both countries.

Many aspects of these resources have had a powerful effect, but only a few will be discussed here. *Principles and Standards for School Mathematics* not only lists standards for math instruction but also articulates the importance of high expectations for *all* students, a coherent curriculum, teacher understanding of the curriculum, student understanding of the curriculum, formative assessment, and the potential for technology to support mathematics instruction.

Most recently, *Principles to Actions* (NCTM, 2014) reiterates that a set of standards on paper does not automatically translate to change. It reminds readers that U.S. students are still not scoring at the highest levels compared to students in other countries and that, in general, U.S. students perform well on low-level items on international tests but not as well on items involving higher-level thinking. It also reminds readers that many U.S. teachers work in isolation, without the benefit of collegial or coaching support. It articulates the principle of professionalism in educators as a critical need.

Particularly valuable in this latest NCTM publication is a set of suggested actions to help leaders come closer to achieving NCTM's goals for student math learning. Details will be discussed later in this resource, particularly those involving providing professional development opportunities for teachers, monitoring instructional time, emphasizing "understanding," maintaining a schoolwide culture with high expectations, ensuring resources meet the spirit of the standards, and performing assessment to guide instruction.

In addition, the introduction of the Common Core State Standards in the United States (Council of Chief State School Officers & National Governors Association, 2010), focusing on understanding as well as performing, has changed the focus in many mathematics classrooms. The eight Standards for Mathematical Practice encourage students to dig into the math and not just repeat rote procedures:

1. Make sense of problems and persevere in solving them.
2. Reason abstractly and quantitatively.
3. Construct viable arguments and critique the reasoning of others.
4. Model with mathematics.
5. Use appropriate tools strategically.
6. Attend to precision.

7. Look for and make use of structure.
8. Look for and express regularity in repeated reasoning.

Examples of student work based on these processes can be found in many resources (Small, 2017c).

What Does Great Math in a School Look Like?

Fleshing out the ideas presented by NCTM helps us imagine what we would see in a school where great math is happening.

- Instructional tasks combine significant attention to mathematical thinking with more straightforward skill and application questions. For example, students might figure out ways to show $20 using different combinations of coins, instead of only figuring out the specific change from $20 for a $4.59 purchase.
- Math work is visible around the school, whether it is a tower of blocks showing a pattern, a hundreds chart used for games out on the playground, or student-created videos of math problems or solutions shared with the school community and parents (see Figure 1.1).
- Virtually every teacher who teaches math (and even some who do not teach math) is interested in mathematics instruction and is aware of some of the bigger ideas across the grades. Virtually *every* student feels confident trying math tasks as well; after students have worked on these activities, you will often hear, "Can we do another one?"

Notice that math scores are not the major consideration. High scores do not necessarily mean that math understanding is at a deep enough level. Too many students can successfully repeat what a teacher has shown them but hit a wall when any element of a problem or situation changes. Thus, their work might look good at first glance, but the learning might not be solid enough to transfer to a new or more complex situation.

Figure 1.1

FIGURE CUP PATTERN

PLAYGROUND HUNDRED CHART

HOW MANY CHICKENS?

Deep down, it is the curiosity about math and the ability to solve problems—not just the ability to calculate and solve equations—that we are looking for in our students. Often principals believe that great change is happening in the school without much data to back it up. They usually notice one good thing that is happening somewhere and forget the rest. Figure 1.2 lists specific practices that administrators should look and listen for in their schools to determine whether great math is taking place.

What About Meeting the Standards?

Students may obtain good test scores, but, in the end, students are meeting the standards only if the tests reflect both the letter and the spirit of those standards. Consider each of the following pairs of activities; they both relate to the same standard, but only one adequately addresses it.

Grade 3 Standard: Interpret products of whole numbers; e.g., interpret 5×7 as the total number of objects in 5 groups of 7 objects each.

A: How much is 5×7?

B: Draw a picture to show what 5×7 looks like.

Task B addresses the standard, but Task A does not. Task A asks for a specific number, not for an understanding of the operation; the student could simply recite the fact from memory rather than apply the concept of multiplication.

Grade 6 Standard: Determine, with and without computation, whether a quantity is increased or decreased when multiplied by a fraction, including values greater than or less than one.

A: What is $\frac{2}{3} \times \frac{6}{5}$?

B: Tell or show what number you might multiply by $\frac{6}{5}$ to obtain a product that is a little more than $\frac{3}{5}$.

Figure 1.2 | **What Does Great Math Look Like and Sound Like?**

I see great math when . . .	What I won't see	How will I know?
Students and teachers look forward to math time.	Math is students' least favorite time in the day.	Directly ask students whether they are excited about math class or are not looking forward to it.
Math is focused on solving problems and thinking and not just recall of procedures.	Children are filling in worksheets.	Worksheets might be assigned occasionally, but their use is not widespread within a class or across classes. You might ask students to tell you about the most recent math problem they solved. It is only a problem if they didn't already know how to do it, so you need to ask what made it something they had to think about.
Students are communicating mathematically during a large proportion of their learning time and are engaged in the assigned activities.	There is a lot of student silence or lack of engagement.	You might randomly listen to groups of students in different classrooms and evaluate how rich their math conversations are.
Students regularly choose and use visual representations and manipulatives to support and demonstrate their mathematical thinking.	Visual tools are rarely used and are often used in an algorithmic or procedural way, not for solving problems or thinking mathematically.	Visual tools include manipulatives and representations such as number lines, hundred charts, and so on. Students use tools even when not instructed to do so. You might ask students why they chose a particular tool for a math problem. They should not always be choosing because "that's the one the teacher said to use"; they should have their own reasons.
The tools and strategies become more sophisticated as students advance through the grade levels.	New tools and strategies are not introduced as students move from grade to grade.	This can be observed by looking at student work on similar problems at different grade levels. There should be meaningful changes both in the tools and the ways in which they are used.
Students regularly ask substantive questions of their peers and of the teacher.	Basic questions that maintain a focus on organizational or procedural aspects of a task are the only ones students ask.	You might listen for the first five or last five questions in different classrooms on different days and observe the proportion of questions that are substantive.

Figure 1.2 | **What Does Great Math Look Like and Sound Like?** *(continued)*

I see great math when . . .	What I won't see	How will I know?
Students expect daily math challenges and rise to them.	Students expect to be shown everything and do not want to be asked to figure things out.	You might ask students if they are usually shown how to do a problem before they are asked to do it or whether they often figure things out themselves.
Nobody talks about math being "hard."	Teachers openly talk about finding math hard or not liking math to their students or colleagues.	Ask students if they have ever heard other people say that math is hard and ask who those other people were.

Task B addresses the spirit of the standard better than Task A. Although the student will find a solution with Task A, it does not ask the student to determine whether a quantity was increased. On the other hand, Task B allows for students to either use calculations or reasoning to consider how the value of a factor changes when multiplied by a particular fraction.

High School Standard: Describe the relationship between the linear factors of quadratic expressions and the zeros of their associated quadratic functions.

A: What are the zeros of $y = x^2 - 5x + 6$?

B: How does knowing that $x^2 - 5x + 6 = (x - 3)(x - 2)$ help you sketch the graph of $y = x^2 - 5x + 6$?

Task B addresses the standard better than Task A because it focuses on the required connection between factors and zeros.

Although state or provincial tests are usually constructed with the standards in mind, often teacher-created tests are a mix of questions that adequately test the standards and those that do not.

What Kinds of Questions Best Meet the Standards?

Questions should attend to the practice, or process, standards as well as the content standards. The standards for practice, or process, need to be assessed regularly and provide important information about whether standards are being met.

For example, one important standard in many states, modeled after the Common Core State Standards, is that students can "construct viable arguments and critique the arguments of others" (Council of Chief State School Officers & National Governors Association, 2010). This ability needs to be assessed. Therefore, a principal might look at teacher tests, quizzes, or assignments for items like these.

- **Grade 2:** "Why can't the answer to 2□ + 2□ possibly be in the 60s?"
- **Grade 7:** "When you multiply two numbers, the answer is less than when you divide them. What numbers can they be? Which can't they be?"
- **High school:** "You multiply two irrational numbers that are not the same. How could the answer be rational?"

Questions should ask for explanations and not just answers.

- **Grade 2:** "How do you know that 4□ + 5□ has to be almost 100 *without finding the answer*?"
- **Grade 4:** "How do you know that $\frac{3}{5} - \frac{1}{3}$ must be more than $1 - \frac{1}{10}$ *without finding the answers*?" The students should realize that $\frac{1}{10}$ and 1 are far apart compared to $\frac{1}{3}$ and $\frac{3}{5}$, which are both between $\frac{1}{10}$ and 1.

Questions should ask students to draw a visual to show an idea. For example, a prompt like "Draw a picture to show why $\sqrt{18}$ equals $3\sqrt{2}$" is valuable. A student who realizes that the two right triangles in Figure 1.3 must be similar shows an understanding of why $\sqrt{18}$ equals $3\sqrt{2}$.

Questions should ask students to make connections.

- **Grade 4:** A teacher might ask, "How is subtracting $\frac{2}{3} - \frac{1}{5}$ like subtracting 10 – 3? How is it different?"
- **High school:** A teacher might ask, "How are the graphs for $y = 2^x$ and $y = 2^{2x}$ alike? How are they different?"

Questions should be open-ended enough to allow students to show their interpretation of ideas.

- **Grade 1:** "Do you think 15 is more like 10 or more like 20? Why?"
- **Grade 6:** "The greatest common factor of two numbers is one third of one of them. What do you know about those numbers that can help you tell what they might be?"

Figure 1.3 | **Pictures Can Show Ideas**

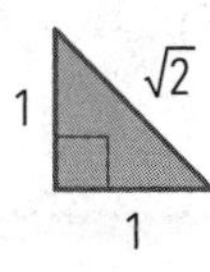

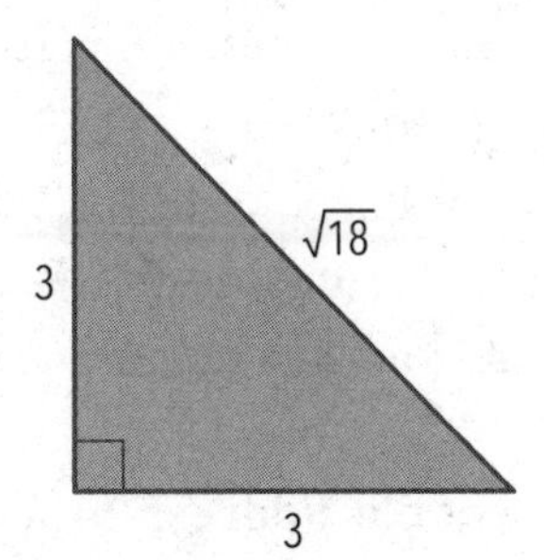

What Kinds of Resources Best Meet the Standards?

To meet standards, teachers need to use appropriate resources, whether print or digital. A principal must determine whether the resources that are requested and used by teachers are ones that support the spirit of the standards. What would a principal look for when choosing resources?

The resource should focus as much on understanding and thinking as it does on skill development. There are many examples of resources (computer programs, texts, or supplementary resources) that say they meet standards but actually address only some of the content, and certainly not the spirit, of those standards.

- **Grade 2:** In a lesson on subtraction, students should have repeated opportunities to see the addition inherent in any subtraction situation. They might be asked, for example, how they could use addition to determine 57 – 12.
- **High school:** In a lesson on quadratic equations, students might be asked why most measurement situations involving quadratics focus on area and not on volume or length.

The resource should not regularly model tasks such as how to perform computations, find the rule for a pattern, or count the number of diagonals in a shape. If the resource models for students too much, it is not adequately calling for problem solving. It should leave room for students to figure things out for themselves.

The resource should speak to the teacher as a professional, and the teacher component, at a minimum, should clarify for teachers why certain directions or actions are taken and encourage teachers to reflect on those rationales.

The resource should address differentiation, providing meaningful and thoughtful extensions for strong students and meaningful and thoughtful scaffolds for struggling students (not just easier questions).

The resource should focus students on building connections between ideas they have learned and not treat each math topic as a separate entity.

The value of a resource is not related to the medium (print or digital). There are print textbooks that ask thoughtful questions (e.g., "Use a different strategy to figure out 13 - 8 than to figure out 9 - 2"), and there are digital resources that focus on straightforward, rote procedures.

Why Aren't We Seeing as Much of This Rosy Picture as We Should?

There is a *long* history of concern about math instruction being the domain of a small number of interested teachers, whether they consist of the math department in a secondary school or a few keen teachers in an elementary school. Elementary school teachers tell us that they have very little math in their backgrounds and often lack a deep understanding of the curriculum they are delivering; for this reason, they "bow out" of trying to change mathematics instruction. Often they say they feel no pressure to try to improve their math instruction because many of their principals also stay out of math discussions due to their own discomfort in that domain. Most principals readily admit that they defer to their math departments. In addition, many high school teachers today are the products of a more traditional approach to math, so their own experiences as students create an extra hurdle as they try newer approaches in their own classrooms.

With the advent of more current curricula and the desire to improve the math performance of our students on international assessments, these practices will no longer suffice. Underlying decisions that are or are not being made about how mathematics should be taught reflect strong differences in belief about what math really is. It would be interesting for a principal to try this poll in Figure 1.4 with staff (perhaps anonymously) to see just how different their beliefs are; it could be the start of a valuable staff discussion.

Figure 1.4 | **Belief Survey**

Circle your response to each statement.			
The big focus in math instruction should be on application.			
Strongly agree	Agree	Disagree	Strongly disagree
Covering curriculum matters more than making the math interesting to students.			
Strongly agree	Agree	Disagree	Strongly disagree
Developing positive attitudes toward math is challenging.			
Strongly agree	Agree	Disagree	Strongly disagree
Math is mostly about developing computational and algebraic skills.			
Strongly agree	Agree	Disagree	Strongly disagree
Most of my teaching time in math should focus on problem solving.			
Strongly agree	Agree	Disagree	Strongly disagree
Math is more about procedures than concepts.			
Strongly agree	Agree	Disagree	Strongly disagree
Most math teaching should be direct instruction, with some guided and some exploratory work.			
Strongly agree	Agree	Disagree	Strongly disagree
The weakest students in my class might improve, but they really can't be expected to achieve at top levels.			
Strongly agree	Agree	Disagree	Strongly disagree
Open-ended problems should be used rarely because they are too hard to grade.			
Strongly agree	Agree	Disagree	Strongly disagree

When teachers have fundamentally different views about the nature of math, it is hard to develop a school culture. However, it remains essential to come to terms with that vision collectively.

Big Picture: What Data Would Tell You Whether Your School Is Great?

There are a variety of ways to obtain data, beyond scores on external tests or teacher marks, that will help you get a pulse on your school. (Some of these will be explored in more depth later in this book.) To use these measures, however, there must be deliberate attention to collecting kinds of data that are often not considered.

Data collected about students reflect teaching practices. Whether the data are taken from individual students or collected from an entire class, student work is a good indicator of what is being taught or emphasized and how the teaching in one class might differ from the teaching in another. Because the principal's focus is on teaching practice, it makes sense to gather data that reflect that practice and show how it has influenced students. The data that you collect can be either quantitative or qualitative.

Examples of Quantitative Data

- Measures of how the work of students at different grade levels compares on calculations or problems (counting/analyzing use of various strategies)
- Measures of the level of depth of student thinking/reasoning
- Measures of how long students persevere on rich problems or on external tests
- Measures on Likert scales of positive mindset or positive attitudes toward math (Dickson, 2011) (perhaps teacher mindset, too)
- Measures on Likert scales of student confidence in their math abilities (perhaps teacher confidence, too) (Hendy, Schorschinksy, & Wade, 2014)
- Measures of the proportion of student talk compared to teacher talk in a given class period
- Measures of the number of substantive questions asked by students during a math class

Examples of Qualitative Data

- Measures of enthusiasm for math
- Measures of curiosity about math
- Measures of student creativity in math

Summing It Up

In this chapter, we showed what a positive math culture looks like in schools that foster students' mathematical thinking in meaningful ways while meeting rigorous standards, and we outlined a variety of indicators of student success beyond student engagement and test scores. High among these are the following:

- Math is focused on solving problems and thinking and not just recall of procedures.
- Students regularly use increasingly sophisticated visual representations to support and demonstrate their mathematical thinking.
- Students expect and rise to daily math challenges. Those traditionally seen as weak students no longer look weak; all students are improving.

We discussed the importance of teacher use of high-quality teaching resources. We also discussed the importance of data as a powerful tool to help administrators determine the strength of the math culture in their schools. The data that can and should be collected to measure math performance include evaluations of both teaching practice and student work, as well as attitudes among staff and students toward mathematics.

This deeper look at positive math culture emphasizes how great math is more than great test scores. It requires much more to give student learning the context, rigor, and depth it needs to make it meaningful. The best first step toward creating this kind of learning in mathematics across your school is with teachers, which is the subject of the next chapter.

2

Laying the Foundation for a Strong Math Culture in Your School: What Are the Critical Steps?

Basing Decisions on Data That Matter

Many administrators rely on anecdotal evidence from their teachers to get a sense of the changes in mathematical practice that have occurred in their schools. However, there are inherent problems with anecdotal data. For example, a teacher may report that she is using manipulatives more in her classroom, but perhaps she only used manipulatives with a few students a few times. In her mind, she interprets this as an extensive use of manipulatives, but this is perceived reality, not actual practice. More importantly, even if the teacher has encouraged increased manipulative use, unless students benefitted from that use or unless students chose the manipulatives themselves, there has been no real change. Ultimately, the evidence must be found in student work, not just in reports about that work. Therefore, the data that principals need to collect should emphasize student outputs.

Using Common Tasks to Build a Whole-School Framework

To focus on schoolwide change, principals need to collect schoolwide data. That means selecting a topic for study that applies to the whole school. Data related to that topic can then be gathered, examined, and analyzed for trends within a classroom or for trends with individual students. For example, the topic of addition and subtraction pertains to virtually any grade K–6, proportional reasoning pertains to any grade 5–8, and algebraic tasks pertain to most grades 9–12. A common task might be built for each of those grade bands, with some variations for different levels (Duff & Tonner, in press).

Attention should be given to the kind of task that makes sense for these common topics. Initially, a leadership team, comprising educators who are comfortable with math, will most likely develop these tasks. This leadership team should involve the principal and a variety of teachers. The nature of the task should flow from what research has suggested matters for a particular topic. For example, if the task is about percentages, students must understand several important concepts:

- A percentage can be represented as a fraction or a decimal.
- A percentage can be modeled without using 100 items.
- Percentages can be less than 1 percent or greater than 100 percent.
- The same number can be a different percentage of different numbers.
- Knowing one percentage of a number automatically gives the other percentages of that number.

The task should, as simply as possible, address some or all these ideas.

Ideally, if possible, the task should be differentiated so that individual students can perform it at their respective levels. In the previous example, some students may struggle with percentages less than 1 or greater than 100; therefore, the common task might include these kinds of percentages but not require mastering them. In this way, we learn what a student *does* know about percentages and not just what the student does *not* know.

Criteria for assessing student work are developed communally based on the type of models, representations, and strategies teachers would expect to see. This is because a focus on mathematical models underlies much of the work in grades K–12. Any score given is not as much a number (or level) as it is a way to provide details about the level of sophistication of the child's work.

The criteria will vary from topic to topic, task to task, but, in general, there should be some common expectations (Duff & Tonner, in press):

- **Students use mathematical models.** The sophistication of the mathematical model provides valuable information about the quality of the student's thinking. For example, if a task involves working with greater numbers, base ten blocks or number lines are effectively employed, rather than counters that each represent one unit.
- **Students can use a variety of models to achieve a result.** For example, if asked to model a computation like $\frac{2}{3} + \frac{1}{5}$ in two different ways, students can perform the task.
- **Diagrams match the thinking described.** If a student is asked to show the meaning of 53 – 17 by adding on to 17, for example, the diagram shows this. It will not show 53 items and removing 17, which is correct but does not demonstrate the required thinking.
- **Where there is a specific answer, students should achieve the required result.**
- **No misconceptions are evident.** For example, when adding fractions, the student does not add the numerators together and the denominators together to determine the numerator and denominator of the sum.

Sample Common Task for Grades 1–5

Choose one of these calculations. Show at least two different ways to find the answer.

- $10 - 3$
- $50 - 29$
- $4.1 - 1.78$
- $\frac{8}{3} - \frac{2}{5}$

Vague terms like "improvement in mathematics" should not be the target. Instead, it is important to evaluate specifics about the number and types of models the students choose, the sophistication of reasoning students use to arrive at mathematical conclusions, and the ability of students to communicate their mathematical reasoning. Some criteria that can be used for analysis are shown in Figure 2.1 (Duff & Tonner, in press).

Students may use a wide variety of models to assist in performing this task:

Number line. Notice the difference between the two number line models in Figure 2.2. The first focuses on subtraction as "take away." Students might simply count back 29; they might also count back 30 and then 1 forward or might count back 20 and then back another 9. Any of these are acceptable strategies.

In the second number line, students think about subtraction as the inverse of addition—a powerful mathematical concept. They count the number of spaces from 29 to 50, possibly by moving forward 1 to arrive at 30 and then another 20 to get to 50. Some students might move from 30 to 51 instead, recognizing that the distance must be the same; this is a powerful mental strategy.

Base ten blocks. Notice that the first approach models the standard algorithm and requires regrouping in Figure 2.3. However, the second

Figure 2.1 | **Analysis Criteria for Subtraction Task**

Subtraction Task

Evidence	Seen	Comments
Correct		
Correct answer		
Appropriate strategies		
Varying strategies		
Different models or representations		
Different meanings of subtraction		
Number line		
Base ten blocks		
Ten frames		
Decimal grids		
Fraction materials (e.g., fraction strips)		
Grids		
Incorrect		
Only "take away" model		
Place value errors		
Misconceptions		
Numerators and denominators subtracted		
Decimals lined up from the right		

Source: From *MathUP School: School Improvement Cycle for Mathematics*, by D. Duff and J. Tonner, in press, Oakville, Ontario, Canada: Rubicon Publishing. Copyright 2018 by Rubicon. Reprinted with permission.

model, which also uses base ten blocks, focuses on subtraction as comparison and requires no regrouping. Instead, the student sees that two tens and one unit are needed to complete the total.

Ten frames. Figure 2.4 models subtraction both in terms of take away, in the first approach, and in terms of missing addend, in the second approach. Both are important for students to meet.

Figure 2.2 | **Number Line for 50 – 29**

It might be found by "jumping back" 29 from 50. The difference is represented by the number at the end of the "jump."

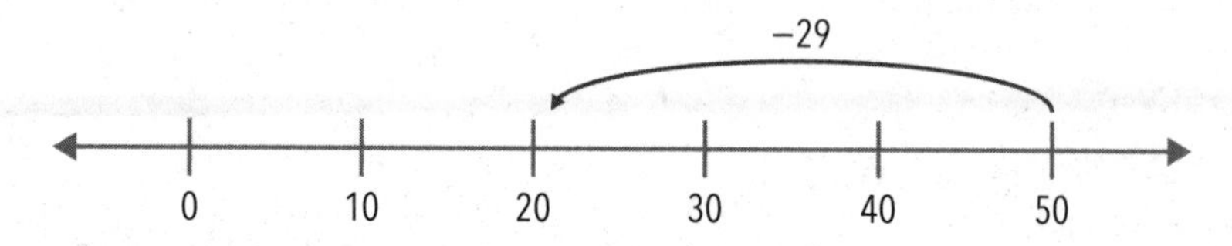

It might also be found by "jumping up" from 29 to 50. This time the difference is the number of spaces moved.

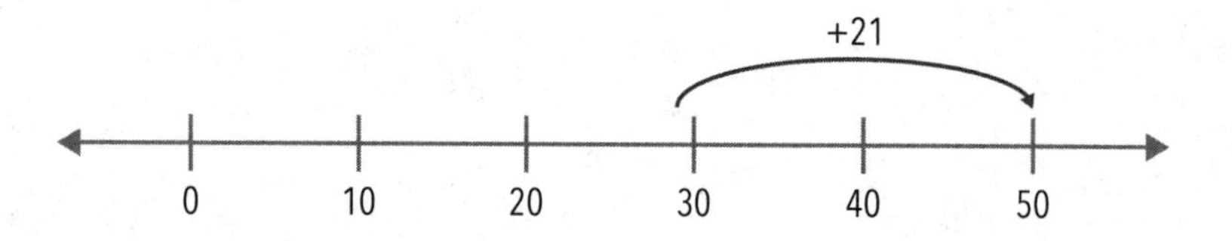

Decimal grids. Figure 2.5 focuses on the inverse relationship between addition and subtraction. The student also realizes that 1.78 is embedded in 4.1, and the difference is simply "the rest." The student might also record this with numbers.

$$
\begin{array}{rl}
4.1 = & 1.78 + 0.02 + 0.2 + 2 + 0.1 \\
-1.78 = & -1.78 \\
\hline
2.32 = & \quad 0 + 0.02 + 0.2 + 2 + 0.1
\end{array}
$$

Fraction strips and grids. Note that in Figure 2.6, two wholes plus $\frac{4}{15}$ is required to make $\frac{2}{5}$ as long as $2\frac{2}{3}$.

Alternately, the student might use grids (see Figure 2.7). This model is based on the concept of area.

Figure 2.3 | **Base Ten Blocks for 50 – 29**

It might show removing 29 from 50 by "trading" one ten for 9 units.

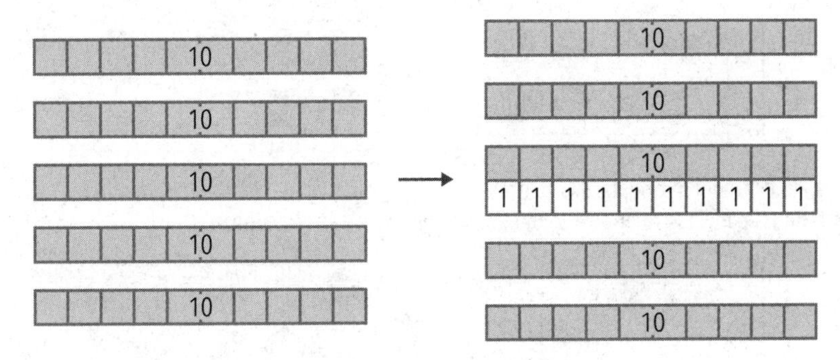

The student then removes 29. This leaves two tens and one unit, or 21.

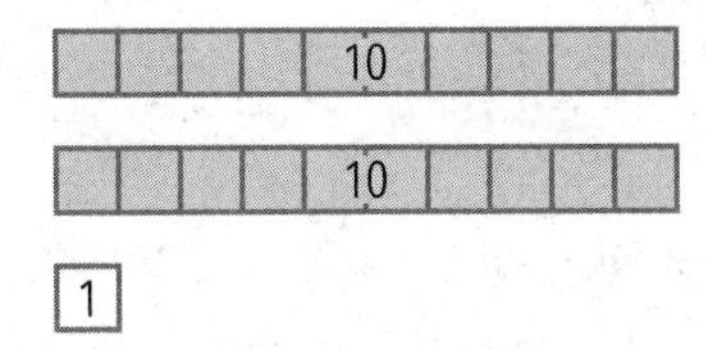

Alternately, the student might show 29 compared to 50, counting how many more are in 50.

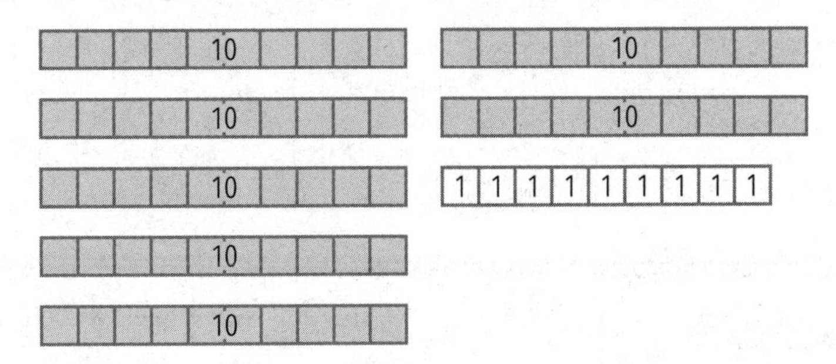

Figure 2.4 | **Ten Frames for 10 – 3**

The student might start with 10 counters and remove three.

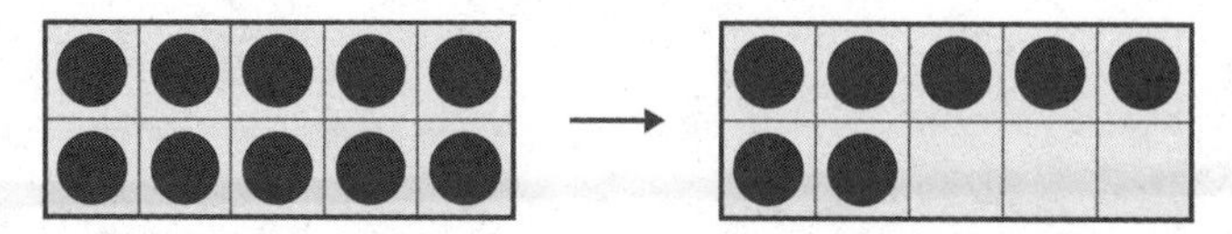

Alternately, the student might start with three counters and count the number needed to make 10, demonstrating the inverse relationship between addition and subtraction.

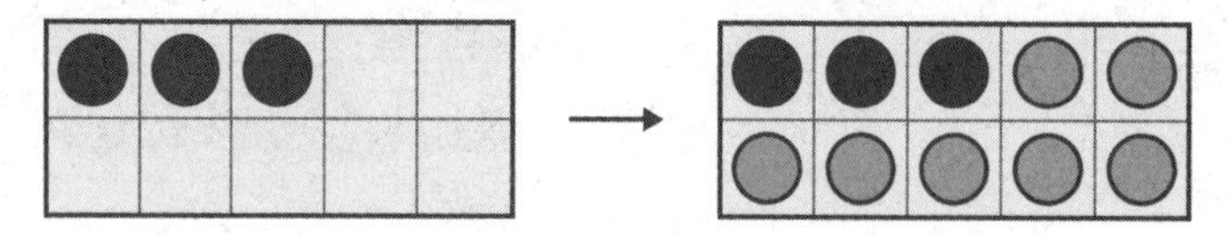

Figure 2.5 | **Decimal Grids for 4.1 – 1.78**

A student might show 1 and 0.78 and then shade in another 0.22 to fill the second grid.

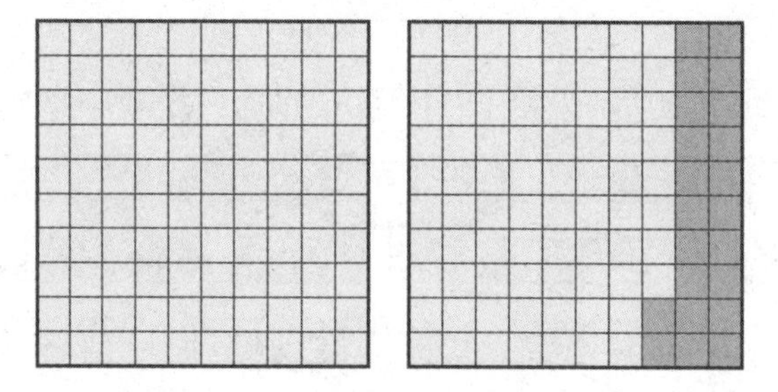

Then the student might add another 2.1 to get a total of 4.1 filled grids.

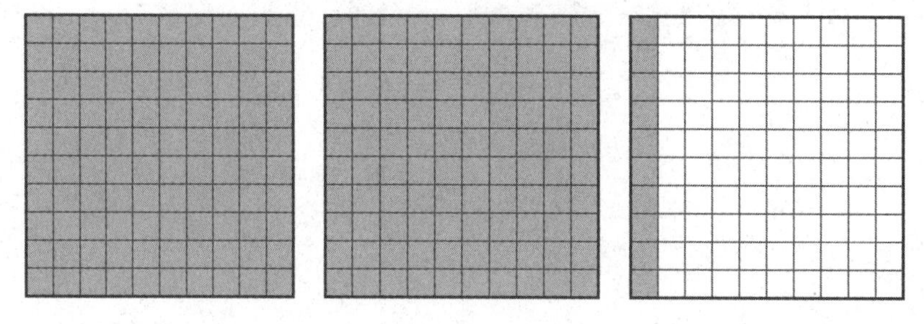

Figure 2.6 | **Fraction Strips for $\frac{8}{3} - \frac{2}{5}$**

A student might use fraction strips to model six thirds (two wholes) and two thirds of another whole. The next step is to determine what must be added to $\frac{2}{5}$ to make $\frac{2}{3}$.

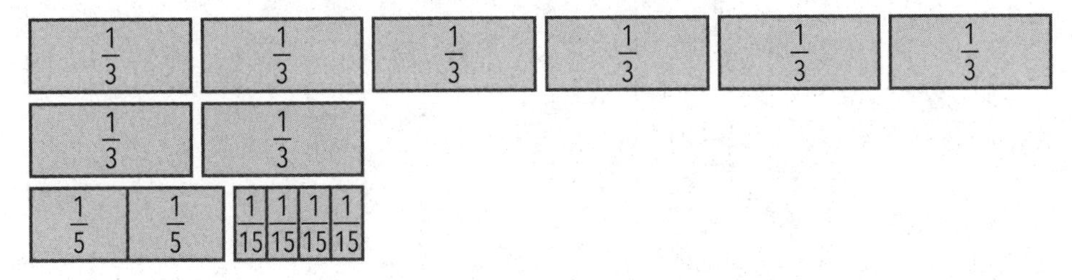

Figure 2.7 | **Grids for $\frac{8}{3} - \frac{2}{5}$**

This student first represented $\frac{8}{3}$ by filling eight columns in the grids. (Each column is one-third of a grid.)

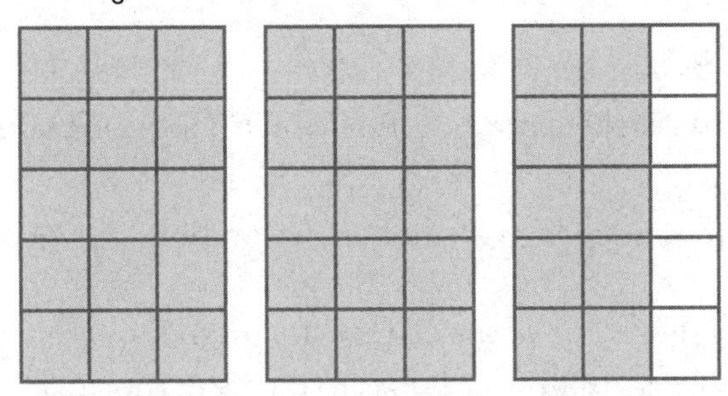

Then she moved colored squares so that two out of the five rows in one grid can be removed.

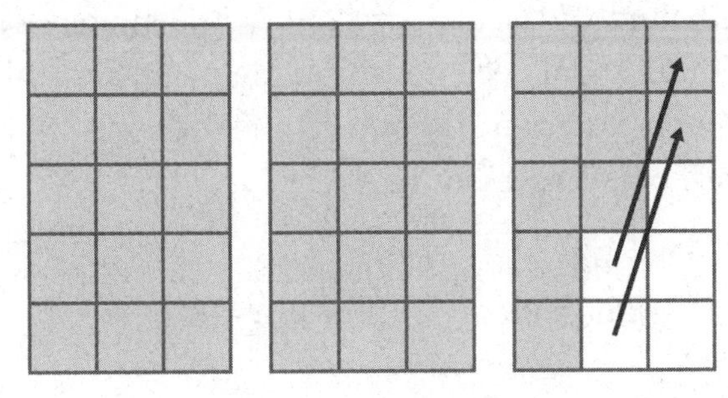

Figure 2.7 | **Grids for $\frac{8}{3} - \frac{2}{5}$** (continued)

Then she removed those two rows (or $\frac{2}{5}$) of the colored squares from the last grid.

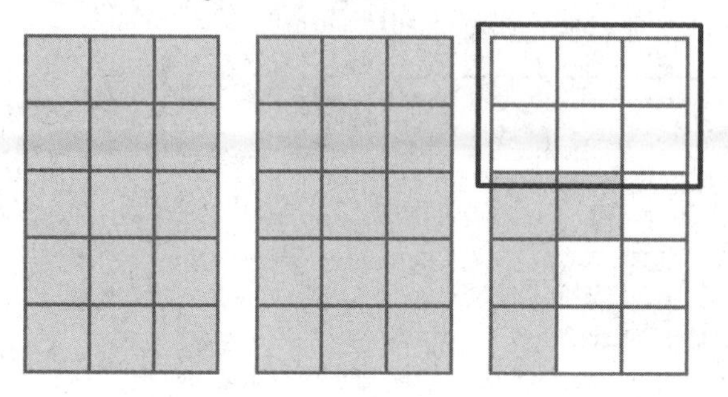

Sample Common Task for Grades 5–8

Choose one of these problems. Show three different ways to solve it. Use at least one model.

- What is 25% of 396?
- If eight batteries cost $3.76, how much should six of those same batteries cost?
- You drive at a speed of 48 mph. How long will it take to go 588 miles?

Figure 2.8 shows some of the criteria for analysis.

Students may use different models to perform this task.

Number line or double number line. There are, of course, many ways to show this with numbers (see Figure 2.9). A student may write 0.25 × 396 but might also think about decomposition. He might also think of the relationship between fractions and percentages, expressing 25 percent as $\frac{1}{4}$ and then thinking of $\frac{1}{4}$ of 396 as dividing 396 by 4.

$$4\overline{)400-4}^{\,100-1} \quad \text{or} \quad 4\overline{)360+36}^{\,90+9} \quad \text{or} \quad 4\overline{)100+100+100+80+16}^{\,25+25+25+20+4}$$

Figure 2.8 | **Analysis Criteria for Proportion Task**

Proportion Task

Evidence	Seen	Comments
Correct		
Correct answer		
Appropriate strategies		
Varying strategies		
Different models or representations		
Relating percent to fractions and decimals		
Number line or double number line		
Tape diagrams		
Graphs		
Hundredth grids		
Decomposition		
Doubling and halving		
Incorrect		
Number lines not scaled accurately		
Labeling/scale on graphs		
Misconceptions		

Source: From *MathUP School: School Improvement Cycle for Mathematics*, by D. Duff and J. Tonner, in press, Oakville, Ontario, Canada: Rubicon Publishing. Copyright 2018 by Rubicon. Reprinted with permission.

Hundredths grid. Teachers might discuss whether the number line or the hundredths grid (see Figure 2.10) is a more sophisticated approach and why. There may not be total agreement on this.

Tape diagrams. The tape diagram (see Figure 2.11) helps the student see that six batteries would cost $1.88 + 0.94, or $2.82.

Graphs. The student can see that it takes a little more than 12 hours (which makes sense because 12 × 48 = 576) in a graph (see Figure 2.12).

Figure 2.9 | **Double Number Line for 25% of 396**

A student might use a double number line, matching 100% to 396.

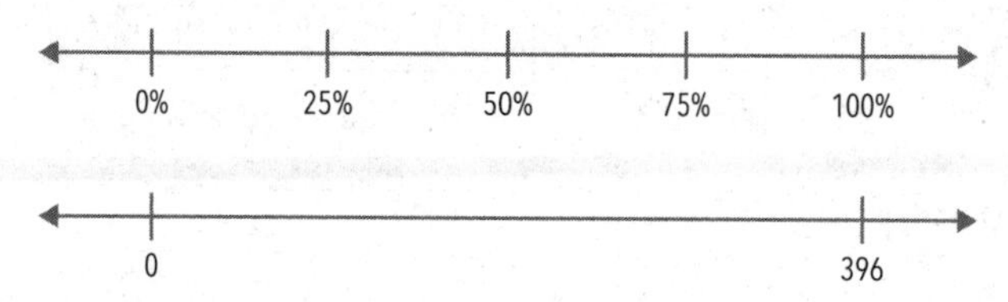

Next, the student finds matches for other benchmarks.

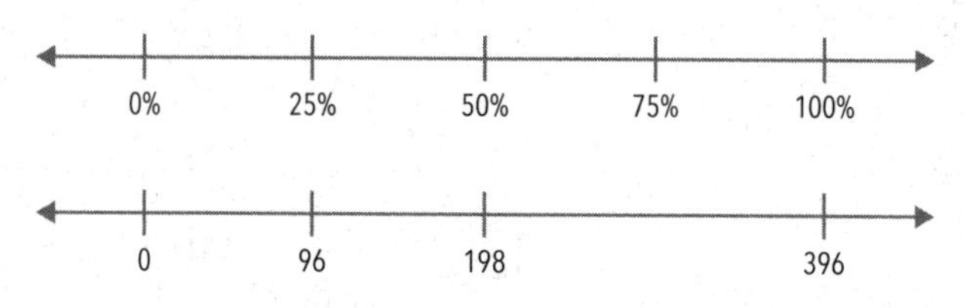

Figure 2.10 | **Hundredths Grid for 25% of 396**

The student decides that the whole grid (or 100% of the grid) is equal to 396. That means each column has the value of 396 ÷ 10, 39.6, and each half-column has the value of 19.8, half of 39.6. She then notices that 25 squares (or 25%) would be equal to 39.6 + 39.6 + 19.8, or 99.

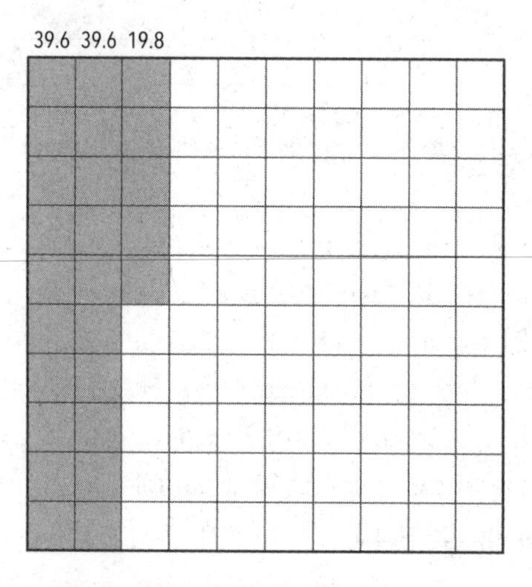

Figure 2.11 | **Tape Diagram for Batteries Problem**

A student might draw tapes to model this problem.

8
$3.76

which is equivalent to

4	4
$1.88	$1.88

or this

4	2	2
$1.88	$0.94	$0.94

Figure 2.12 | **Graph for Mileage Problem**

A student could use a graph to determine how long it would take to travel 588 miles at a speed of 48 mph.

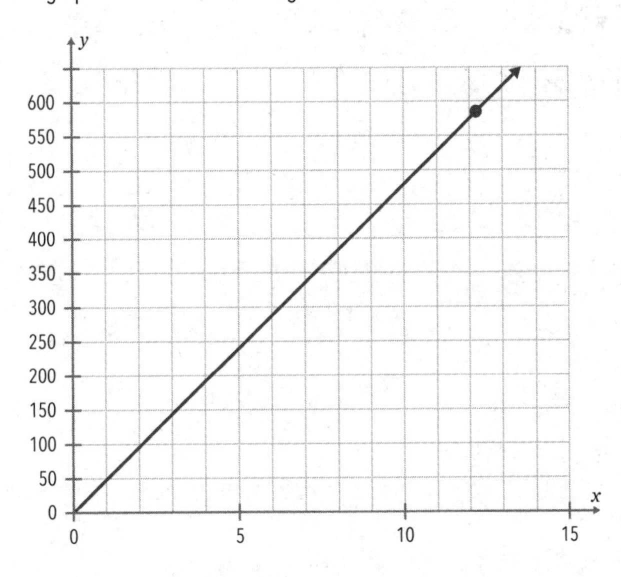

Sample Common Task for Grades 9–12

Choose one of these equations. Show three different ways to determine an answer.

- $\frac{1}{2}x - \frac{2}{3} = \frac{3}{4}x - 2$
- $3x^2 + 19x - 14 = 0$
- $-4x - 17 = 2x + 8$
- $x^2 - 16 = 36$

Some of the criteria for analysis are shown in Figure 2.13.

Figure 2.13 | **Analysis Criteria for Algebra Task**

Algebra Task

Evidence	Seen	Comments
Correct		
Correct solution(s)		
Appropriate strategies		
Varying strategies		
Different models or representations		
Tables of values		
Algebraic manipulation		
Factoring		
Incorrect		
Guess-and-test used, without improved guessing		
Incorrect manipulation of inequalities		
Misconceptions		
Adding to one side of an equation but subtracting from the other		

Figure 2.14 | **Table to Solve $x^2 - 16 = 36$**

x	$x^2 - 16$
0	−16
2	−12
4	0
6	20
8	48

The student realizes the solution is between $x = 6$ and $x = 8$, because 36 is between 20 and 48, and refines the estimates.

x	$x^2 - 16$
7	33
7.2	35.84
7.3	37.29

Different models or representations can be used to perform the task.

Tables of values. The student estimates one value of x to be 7.21 in Figure 2.14. (The other value, of course, would be its inverse, −7.21.)

Graphing. A graph like the one in Figure 2.15 helps students to visually see the solution to the problem.

Algebraic manipulation. To solve $\frac{1}{2}x - \frac{2}{3} = \frac{3}{4}x - 2$, a student might manipulate the equation algebraically.

1. Multiply all terms by 12 to simplify the expression.
   $12\left(\frac{1}{2}x - \frac{2}{3}\right) = 12\left(\frac{3}{4}x - 2\right)$
   $6x - 8 = 9x - 24$
2. Subtract $6x$ from both sides and add 24 to both sides:
   $16 = 3x$

Figure 2.15 | **Graph to Solve $-4x - 17 = 2x + 8$**

The student might graph $y = -4x - 17$ and $y = 2x + 8$ and see where they intersect. At that point, the value of x is approximately -4.2.

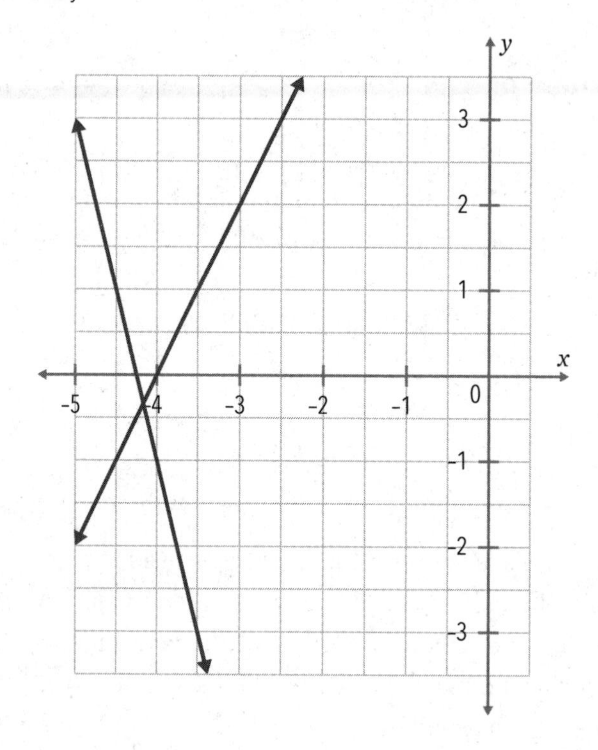

3. Divide both sides by 3.
   $x = \frac{16}{3}$, or $5\frac{1}{3}$

Similarly, to solve $3x^2 + 19x - 14 = 0$, a student might use factoring.

1. Factor the left side of the equation.
   $(3x - 2)(x + 7) = 0$

2. Set both terms equal to 0.
   $3x - 2 = 0$ or $x + 7 = 0$
   $x = \frac{2}{3}$ or -7

When establishing common math tasks for any grade level, the important work is to develop the tasks and then establish the criteria that will be used to evaluate student performance.

Analyzing the Data

The data are then examined by a designated group. Normally that group will involve all in the school who teach math, although, at the secondary level, teachers in related disciplines might also be involved. Generally, this would be an after-school event and would most likely occur a few times during the year.

We believe that a good strategy is to start with some guiding questions for each step of the data analysis process (Duff & Tonner, in press).

1. **Predict.**
   - What do you think "good work" looks like in math?
   - What has our staff been emphasizing?
   - What should look different at the various grade levels?
   - Are there differences in the work in individual teachers' classrooms?

2. **Apply the flip test, having a quick glance through the samples available.**
   - What are our first impressions of the results on the task?
   - Are there common strategies used within a class or grade?
   - Are there obvious misconceptions by class or grade?

- Generally, what do our students know about [topic of task]?
- Are there commonalities in the explanations across the data set?
- Do we notice sophistication by grade across the data set?
- Are students using mathematical models and not just words?

3. **Use the criteria.**
    - Apply the criteria developed for the task to quantify the data.
    - Count occurrences of each of the listed criteria or misconceptions.
4. **Select samples.**
    - Which data pieces are good examples?
    - Which could be used in developing the professional learning piece that will follow?

Some of the points to raise and avoid during these discussions are outlined in the next section.

Making Professional Learning Decisions

After the student work has been seen and discussed, the administrator, with possible support from math coaches or consultants, can decide what sort of professional learning is suitable for the whole group and what sort of work should be developed for individual teachers. It is essential that the professional learning that ensues focus on the student work—in particular, the specific areas where students are having trouble.

It is important not to allow teachers to raise side issues such as class organization, resources, or time constraints. Even if some of the other issues raised are legitimate concerns, they become distractions from the process. For example, if teachers suggest that they have taught a concept and that there is no explanation for poor student performance, the principal should use this as an opportunity to reinforce the power of the assessment criteria and the need for students to show evidence of learning. Often, as staff members use the guiding questions to analyze student work, general ideas emerge that all staff should understand to build continuity and facilitate progression of concepts across the school.

A principal we once worked with used a subtraction audit task (a follow-up task administered to a segment of students to monitor implementation of the process) and the criteria for analysis. She noticed that most of the students understood subtraction only as "take away." When a subtraction problem was posed as a comparison or as inverse addition, the data showed that the students had tremendous difficulty understanding the problem. There was no evidence of understanding the relationship among operations or the relationships between quantities. Given these data, the principal, with help from coaches, consultants, and lead teachers, used these operational relationships and meanings as a module for whole-school professional learning.

Another principal noticed that, in a task where students placed fractions on a number line according to their relative magnitude, most students in a particular class could not place a fraction greater than 1 (e.g., $\frac{4}{3}$). After a discussion with the teacher and a focused analysis of the samples, the principal determined that most of the class did not understand that a fraction could represent a number greater than one. Most students, when questioned while the principal was monitoring, stated that a fraction was "always small" or "just a part of a whole" or that an improper fraction "wasn't really a fraction." The teacher was clear that the topic had been covered as a part of her lessons on fraction rules, but she didn't apply it to the number line. The principal opted to organize some individualized professional learning for the teacher around fractional magnitude and offered the learning to other staff who were interested.

The focus of the whole staff in a school seeking improvement in math needs to be singular. When there are too many goals, it is too easy to "hide";

teachers can always say they are working on *x* when the principal is asking about *y*. In a high school, it might be more about a singular focus for the math department and a commitment from other departments to assist in incorporating numeracy skills and instruction in their disciplines.

Through action research that examines student work on the initial common task and creates and analyzes subsequent audit tasks, teachers develop criteria to access student understandings, get a sense of what good work looks like, and improve their own abilities to document student performance. Teachers improve their self-efficacy; they begin to attribute student improvement or lack of it to their work with the students and not external factors. Research shows that collective teacher self-efficacy has a great effect on student achievement (Hattie, 2012).

The goal is not to force a teacher to admit a need but, instead, to encourage teachers to think about how they want to strengthen their practice. Student work should inform the process, providing individual teachers with more focused support, including suggestions for activities and tasks they can use in their classrooms.

There is value in the whole staff participating in professional learning related to common work, whether it involves addition/subtraction, multiplication/division, proportional reasoning, or algebra. It is rare that deep content knowledge is prevalent among a whole staff in any of these areas, and educators can benefit from sharing their knowledge and experience.

Professional learning should focus on making sure teachers are aware of and comfortable with the range of models students can use in different situations. It might be useful at the end of this learning to provide a summary chart for future reference, such as Figure 2.16 on multiplication strategies.

However, targeted professional learning can be based on teacher request, observations made in a classroom, or principal/teacher conversations.

Figure 2.16 | **Multiplication Chart**

Multiplication means

1. The count based on a set of equal groups
   This is 4 × 3. There are 4 identical sets, with 3 in each set ("4 groups of 3").

- Make sure students realize that sometimes unequal sets can be equalized to show multiplication. For example, this grouping can be rearranged to show 4 × 3.

- Do not suggest that multiplication increases values. It does not when multiplying by 1, for example.

2. The result of repeated addition
   $4 \times 3 = 3 + 3 + 3 + 3$

- Make sure students relate repeated addition to a number line with equal jumps.
  4 × 3

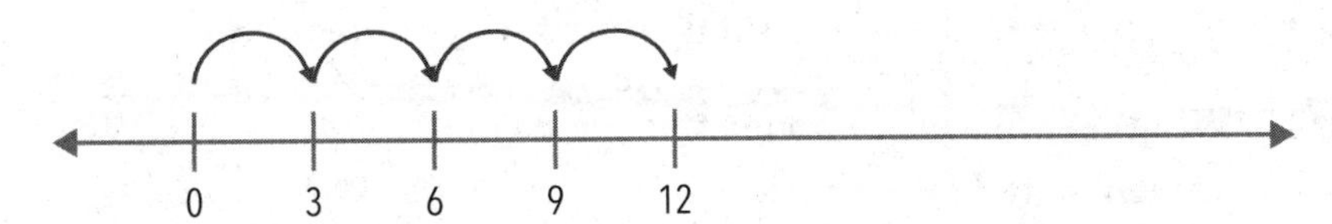

- Make sure students can relate computations. For example, 4 × 3 is also two sets of 2 × 3.

(continued)

Figure 2.16 | **Multiplication Chart** (continued)

3. The area of a rectangle based on its length and width.
   4 × 3 is the number of tiles needed to build a 4 by 3 rectangle.

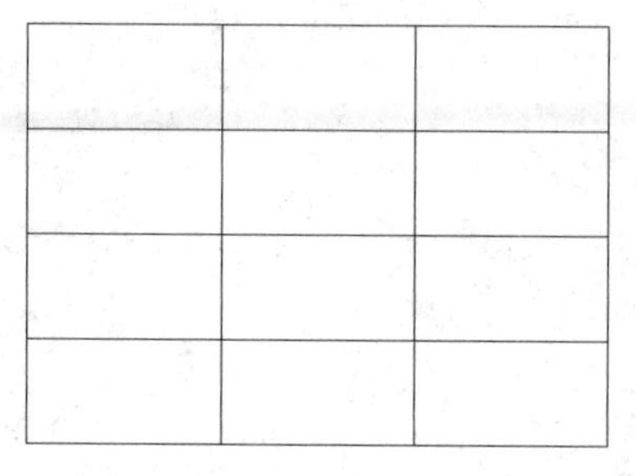

   - Connect this model to the notion of equal groups. Each row or column is a group that is the same size as other groups.
   - This model transfers well to later work with decimals and fractions.
4. A comparison
   4 × 3 can represent an amount that is 4 times as much as a group of 3. There are not really 4 groups of 3, but one can "imagine" the 4 groups of 3.
5. The count of all possible combinations
   4 × 3 is the number of combinations of colors you can make by combining one of 4 shades of blue with one of 3 shades of red. Again, there are 4 "imagined" groups of 3. For each shade of blue, there are 3 combinations created by adding a different shade of red.

This professional learning should be directly related to the analysis and monitoring of the student tasks as identified in the student samples. It is always in response to these questions (Duff & Tonner, in press):

- What do you notice about the use of mathematical models in the student data?
- What do the student samples that were incorrect or incomplete reveal about student understanding?
- What errors or misconceptions in the students' work indicate an insufficient understanding?

These professional discussions need to result in better ability to differentiate instruction, identify student misconceptions and ways to deal with them, describe a year's growth in math, and outline the progression of concepts in math.

Observing change is an important motivator for teachers' continued commitment to professional learning in math and improving practice. Looking at changes in student performance is powerful. Consider the work of two students, Jaden and Lara, in Figure 2.17, based on the given analysis criteria in Figure 2.18 (Duff & Tonner, in press).

As teachers use analysis criteria, their focus for instruction becomes more sophisticated. Work samples show the teacher and the team what is missing in the student thinking and where the teachers need to move for the desired thinking to be represented in the student work. For example, Jaden and Lara might simply have been seen as struggling students who don't really try to do their best. As a result, the differentiation and remediation they needed might not have focused on lagging concepts, which became more apparent with the evidence from the analysis criteria.

Instructional improvement is maintained and developed with audit tasks. These are follow-up tasks administered to a segment of students, whether a single class or several classes, to monitor implementation of the process. They are used to determine whether students have improved their work in an area that was previously deemed insufficient. Assigning several audit tasks during the year provides useful data to evaluate progress. Like the schoolwide or multigrade tasks, these tasks are evaluated with specific criteria related to the tasks. Growth is either seen or not seen; this tells the story (Duff & Tonner, in press).

These focus/audit tasks need not be complex, as demonstrated by these samples.

Figure 2.17 | **Student Work**

Fraction Math Task

Order the following 5 numbers from least to greatest. Use a model and words to support your answer.

1, 2/5, 4/3, 5/6, 1/7

2/5, 4/3, 5/6, 1, 1/7

Check your answer using a **different** model.

Explain why your work was reasonable or what you had to change.

A fraction can be greater than 1.

Is this statement true or false? Use models to prove your answer.

(continued)

Figure 2.17 | **Student Work** (continued)

Fraction Math Task

Order the following 5 numbers from least to greatest. Use a model and words to support your answer.

~~2/5, 4/3, 5/6, 1, 1/7~~

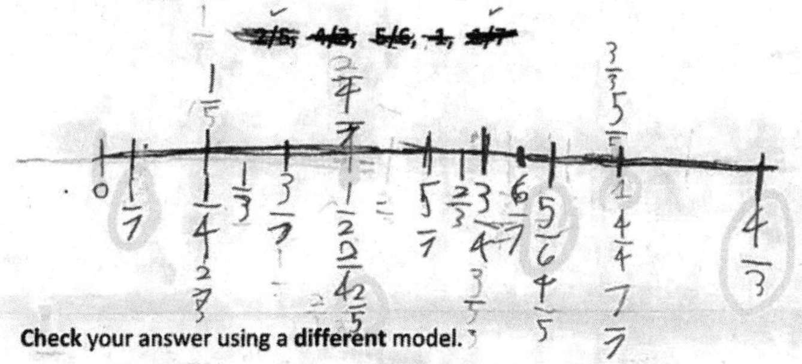

Check your answer using a different model.

Explain why your work was reasonable or what you had to change.

$\frac{1}{7} > \frac{2}{5} > \frac{5}{6} > 1 > \frac{4}{3}$

I used fraction tiles to model the numbers. My answer is correct.

A fraction can be greater than 1.

Is this statement true or false? Use models to prove your answer.

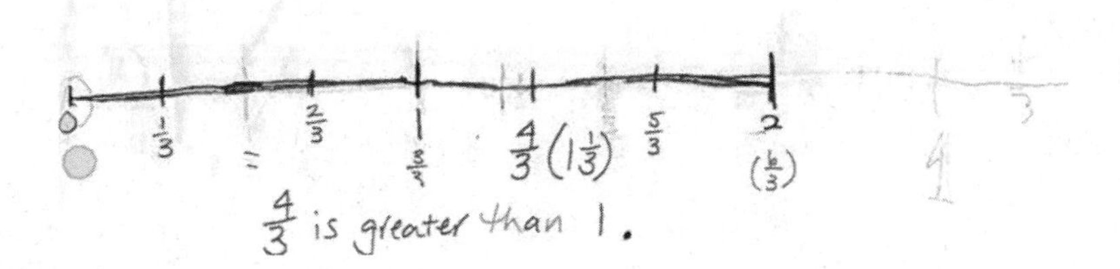

$\frac{4}{3}$ is greater than 1.

(continued)

Figure 2.17 | **Student Work** (continued)

Fraction Math Task January 18th/17

Order the following 5 numbers from least to greatest. Use a model and words to support your answer.

2/5, 4/3, 5/6, 1, 1/7

?

Check your answer using a different model.

Explain why your work was reasonable or what you had to change.

?

A fraction can be greater than 1.

Is this statement true or false? Use models to prove your answer.

I think a fraction can be greater than 1 because you can have numbers like $1\frac{1}{2}$ or $3\frac{2}{8}$.

(continued)

Figure 2.17 | **Student Work** (continued)

Fraction Math Task

Order the following 5 numbers from least to greatest. Use a model and words to support your answer.

2/5, 4/3, 5/6, 1, 1/7

$\frac{1}{7}, \frac{2}{5}, \frac{5}{6}, \frac{1}{1}, \frac{4}{3}.$ ✓

$\frac{1}{7}$ Has smaller pieces than, $\frac{5}{6}$ Is almost whole, $\frac{1}{1}$ Is exactly whole, $\frac{4}{3}$ Is like $1\frac{1}{3}$

explanation do you have pic

Check your answer using a **different** model.

Explain why your work was reasonable or what you had to change.

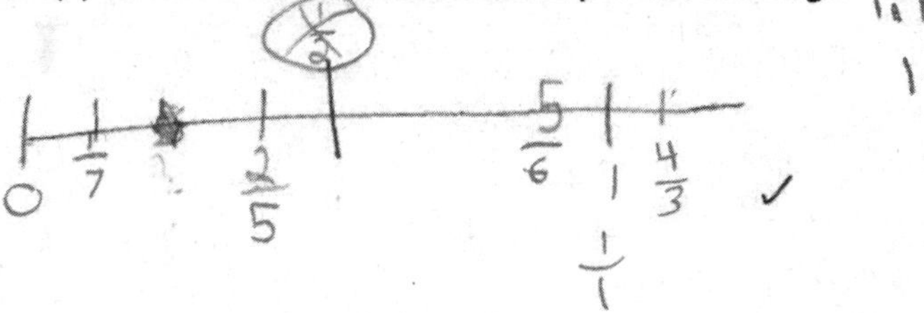

A fraction can be greater than 1.

Is this statement true or false? Use models to prove your answer.

True; example(s). Just gave examples no model

$2\frac{1}{3}$, $1\frac{6}{8}$, $\frac{2}{1}$

Source: From *MathUP School: School Improvement Cycle for Mathematics*, by D. Duff and J. Tonner, in press, Oakville, Ontario, Canada: Rubicon Publishing.

Figure 2.18 | **Analysis Criteria for Fraction Magnitude Task**

Fraction Magnitude Task

Use of Models	Seen	Comments
Correct		
Circle–area models		
Rectangle–area models		
Linear models–number line		
Numerator/denominator comparison		
Fraction as division–decimals, percentages		
Equivalent fraction model		
Numerically ordered–first choice representation		
Fractions of a set		
Ordering by benchmarks		
Incorrect		
Order/no order		
Model–no order		
No understanding of fractions		
Misconceptions		
Inaccurate area–different wholes or sizes		
Linear magnitude not proportional		
$\frac{4}{3}$ as $\frac{3}{4}$ –not understanding fractions greater than 1		
Incorrect model (graph, counters, tallies)		
Incorrect partitioning (e.g., horizontal sections in circles)		
Correct order–no models		
Incorrect use of models or no models		
Ordering by size of numerator		
Incorrect scale on number line		
No second representation		

Source: From *MathUP School: School Improvement Cycle for Mathematics*, by D. Duff and J. Tonner, in press, Oakville, Ontario, Canada: Rubicon Publishing. Copyright 2018 by Rubicon. Reprinted with permission.

Primary Grades

1. Choose two numbers that are 3 apart. Do it again with different numbers.
2. Choose two numbers that are 5 apart. Do it again with different numbers.
3. Choose two numbers where the difference is 10. Do it again with different numbers.
4. Choose two numbers where the difference is 20. Do it again with different numbers.
5. Choose one of your answers to question 1 or 2 and write three story problems using those numbers to show the different meanings of subtraction.
6. Choose one of your answers to question 3 or 4 and write three story problems using those numbers to show the different meanings of subtraction.

(Note to readers: The different meanings of subtraction have been shown previously in various figures: take away, missing addend, and comparison.)

Grades 3–5

1. Write three fractions for each group.
    - Less than $\frac{1}{2}$
    - Between $\frac{1}{4}$ and $\frac{1}{2}$
    - Between $\frac{1}{2}$ and $\frac{3}{4}$

- Between $\frac{2}{3}$ and 1
- Greater than 1

2. Place one fraction from each group on a number line.

Grades 6–8

The number 15 is different percentages of different numbers.

1. Fill in the blanks in many different ways.
   15 is _____ % of ______.
2. Show why three of your answers are correct.

(Note to readers: It is often better not to say how many ways to encourage some students to do even more, if they wish.)

Grades 9–12

1. Create a number of equations where at least one solution for x is -2.5.
2. Try to make the equations as different as possible. Tell how they are different.
3. Tell how you know the solutions include -2.5.

Looking for Change

Early student work can be compared to later work for evidence of improvement. Later work will use identical or similar audit tasks (see Figure 2.19).

Figure 2.19 | **Similar Tasks**

Early Task	Later, Slightly Modified, More Rigorous Task
Choose two numbers with a sum that is double their difference. (Possible values are 3 and 1, 30 and 10, $\frac{3}{4}$ and $\frac{1}{4}$, etc.)	Choose two numbers with a sum that is 20 more than their difference. (Possible values are 4 and 10, 20 and 10, 6 and 10, −1 and 10, etc.)
Name four different numbers *A*, *B*, *C*, and *D* such that $\frac{A}{B}$ is just slightly less than $\frac{C}{D}$. Place each fraction on a number line. (Possible values are $\frac{2}{3}$ and $\frac{4}{5}$, $\frac{5}{9}$ and $\frac{11}{18}$, $\frac{5}{12}$ and $\frac{11}{24}$, etc.)	Name three different numbers *A*, *B*, and *C* such that $\frac{A}{B}$ is just slightly less than $\frac{B}{C}$. (Possible values are $\frac{3}{4}$ and $\frac{4}{5}$, $\frac{3}{8}$ and $\frac{8}{20}$, $\frac{3}{10}$ and $\frac{10}{30}$, etc.)
The product of two integers is −24 and their sum is more than 3. What could they be? (Possible values are 8 and −3, 24 and −1, and 12 and −2.)	The product of two integers is −36. What could their sum be? (Possible values are −35, −16, −9, −5, 0, 5, 16, 9, and 35.)
You solved a quadratic equation, and the roots were three spaces apart on a number line. What could the quadratic equation have been? (Possible equations are $x^2 - 9x + 18 = 0$, $x^2 + 5x + 4 = 0$, and $x^2 - x - 2 = 0$.)	You solved a quadratic equation. One root was positive, and the other was negative. What could the quadratic equation have been? (Possible equations are $x^2 - x - 6 = 0$, $x^2 - 2x - 15 = 0$, and $x^2 - 100 = 0$.)

Summing It Up

Using common multigrade or schoolwide tasks to develop criteria for high-quality student work is key to accurate and consistent assessment of student outcomes in math. The data gained from these tasks aid in developing professional learning opportunities to help staff improve their practice in teaching mathematics. We will have more discussion about data collection, analysis, and application later in this book, but the next step in shifting your school's math culture lies in overcoming resistance and getting buy-in, the subject of the next chapter.

3 Overcoming Resistance and Getting Buy-In

This chapter discusses actions you can take to deal with potential impediments to moving a school forward.

Dealing with Your Own Discomfort with Math

To build a strong math culture, school leadership must lead the charge. However, principals must start with themselves. They cannot lead the way effectively unless they recognize and address any discomfort that they themselves have with math. Often discomfort leads to avoidance. Our experience is that many elementary or middle school principals put an enormous amount of energy and resources into literacy despite school data that show that math is the problem. As leaders, we tend to work on those areas in which we feel competent to provoke change.

A principal we once worked with told us the following:

> "Before this year, I never led a numeracy initiative in my school. It was always incidental or district-driven. As I look back, I now understand that I avoided mathematics improvement from fear that I didn't have the answers or I couldn't recognize what growth looked like in the classrooms. With literacy, the data were readily apparent. I could analyze writing scores or phonological data to find areas of need and growth. I could also participate confidently with teachers and model some of the techniques with my students. I felt in charge of the initiative. I have come to realize, however, that learning alongside my staff, being vulnerable, and engaging my leadership skills had a more powerful effect on the change. Others on staff felt empowered to take more risks and share their learning, as evidenced in the student work."

Principals have both the right and the need to obtain additional support to help them lead. This might include math leads in a school, consultants from the school district, or coaches. These individuals can play an important role in *supporting* the principal, but cannot *take the place of* the principal as the leading change agent.

The principal can show leadership in becoming more knowledgeable in math by visibly reading or working in professional learning situations with the staff. When talking about improving mathematical performance at their schools, one principal shared with us a written goal: "Math this year: Relish my own discomfort and force myself to play with the models. Co-teach, engage with students."

Another principal told us, "I've learned that staff are all at different places. I've learned that we must *all* be invested in the math learning journey in learning new ways of doing things all the time. I've learned to learn alongside others and have others who are knowledgeable to keep my enthusiasm high!"

A third principal shared her initial discomfort with math: "Last year my initial comfort in leading any mathematics change initiative was nonexistent. I was one of those people who said I reached my competency in mathematics at grade 11, and I never wanted to take math again. I was not comfortable with math, and I had no idea how to move the school forward in numeracy."

When asked what she did about her discomfort, she shared,

> "My main focus was to find knowledgeable others that I could access. We had access to an instructional coach, who was pivotal in helping me understand the math that we needed. I also learned about a framework we could use to move our school forward. I worked a lot with our coach and staff lead teachers to help me not only lead appropriately but also to learn together. Having a math improvement cycle, which is a framework I could follow with the assistance of a knowledgeable other, gave me a system to follow that allowed me to comfortably move forward and learn with my teachers. The system consisted of a whole-school common task, followed by analysis of the task data with analysis criteria. From there we would deliver PD [professional development] to our teachers with the coach and the lead teachers as a team. The PD was targeted toward the lagging concepts. Finally, I would monitor the new learning to observe how it showed up in the classroom. I also used focus tasks to see if the new learning reached beyond the teacher and into the student work."

When asked how her school was faring at the end of the year (a school that had been struggling at the start of the year), she said, "We have become a flagship school! From being known as a low-performing school, where the students were not doing well academically, we have become a flagship school where we have visits from other principals in our board and visitors from other district boards throughout the province. Personnel from education organizations have expressed interest in visiting our school, speaking to our teachers, and looking at our student work."

She then shared,

> "My first move was to actually learn the math myself. For example, last September, I did not know there were three meanings of subtraction and that these meanings could connect to other big ideas, like constant difference. Now these ideas are an integral part of my vocabulary and what I look for in classrooms. Once I began to know the math myself, I could confidently work with my staff as we learn the math and move forward. We also began to understand what to look for in the student work and how to develop questions that would deepen not only the thinking of the teachers but the thinking of the students as well.
>
> "Understanding the math myself, networking with colleagues, and having a framework to use within my school have made an amazing difference in how both the teachers and the students see themselves. Some of my students now think of themselves as mathematicians. They *want* to do math. Math was actually considered a favorite subject this year by many students, whereas, in previous years, no one would have said they enjoyed math, let alone it being their favorite subject."

Another principal said,

> "My advice to a principal just starting their journey toward math school improvement is to make sure that you have someone who is knowledgeable and can help you with planning the cycle for math improvement in your building. Also, take the time to learn the math yourself. What was hugely important for me was to bring the learning back and monitor what goes on in the classroom. Speaking of monitoring, my advice is to make sure that your monitoring is always intentional. Use what you've learned from the student work and from the analysis criteria to go into classrooms and observe student talk, teacher-student talk, and solutions to problems. This work will help you plan next steps with your whole staff so the school is on the same page with the new learning."

Building Support

Initially, a principal needs early adopters, or champions, to support the work. There are several foundational steps that should be considered when initiating the work:

1. The goal is to build a critical mass to help develop a new culture into which others can assimilate. Therefore, it is important to start with teachers strong in math; these should be the people other staff members would turn to for support. Let this initial team represent the whole school. They can assist with providing professional learning support, helping the principal shape direction, and analyzing student work.
2. To minimize the belief that teachers are being judged for their performance, make it clear to staff that your focus will remain on supporting their efforts to enhance student performance, looking for evidence in student work. It is never about how hard a teacher works, but instead about whether student performance is enhanced and how principals support teachers in their efforts to do so. Principals might selectively provide professional reading or share something relevant from another classroom to personalize their support of teachers. Principals might share articles or ideas or tweets with a larger group and then have staff discussions about that material. These practices send the message that the principal is both supporting and monitoring for student growth, not strictly judging teacher performance.
3. Make sure the staff works as a team to complete the early work in the process, which focuses on developing analysis criteria for tasks. Different levels of sophistication within and across grades should also be created. Examples of good student work need to keep coming to the table to remind teachers of what is possible.

Exemplars should be kept and catalogued for future use when analyzing student growth across the school over time.

4. Directly involve students. Principals can have students come to them to show their work. Principals can offer feedback or can follow up with students later. They can use public announcements to acknowledge good work by letting students talk about their work or can ask teachers to share some of this work in staff meetings. They can send students from one class to another class to share their work, especially if a student goes to a higher-grade classroom and "impresses" the older students.

We once worked with a principal who made great use of the PA system to build support. Each day, the principal highlighted an example of student work from the previous day's class visits or hallway conversations. The principal intentionally connected the announcement with what was targeted in the professional learning or announced a theme for the day, such as Number Line Monday or Decomposition Day. In a 15- to 30-second segment, the principal would ask a student how to decompose a certain number, what concept he had just discovered, or how she had used something she learned to connect to another area of mathematics. It became the student thinking segment on the announcements, and students were constantly vying for "announcement-worthy" mathematical ideas. Hearing the enthusiastic voices of students and the principal speaking about mathematics and urging others to do the same had a powerful effect on the culture of learning.

Getting Teacher Buy-In

Ironically, resistance can be useful; it can serve as a barometer of change as the level of resistance dissipates. Once resistance is acknowledged, it can help change direction. Overcoming resistance requires five tiers of trust (see Figure 3.1):

- Staff trust in the leader.
- Staff trust in the vision.
- Principal trust in *all* staff (not just champions) to learn and grow.
- Principal and staff trust in students, believing that all students can learn new ideas.
- Student trust in the staff and principal not to be judgmental and to foster openness and risk-taking.

Figure 3.1 | **Trust Links**

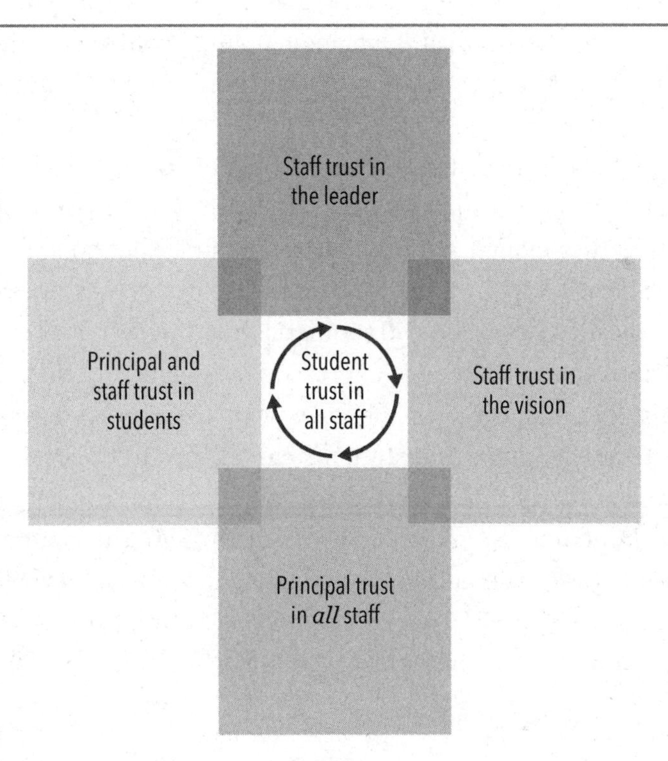

At the start, there must be a focus for change for *all*. If the principal is satisfied with the involvement of just a few champions, it models a lack of trust that *everyone* can change. Even if the champions are the first focus, they can't be the only focus.

The reasons for change must be articulated to engender commitment. This is accomplished using data, testimonials, anecdotes, and other resources. Data showing that students are not where they need to be lead to a sense of urgency, which helps move teachers along. Initial champions can be useful in helping to gather those data quickly and convince colleagues of the need for change.

The principal must continually work alongside teachers both to support and to monitor. Support will initially reduce resistance, but then monitoring is essential for moving forward and ensuring that change efforts are working. Staff, along with the principal, must continually develop and evolve audit tasks to ensure change can be monitored and evaluated. This will be discussed further in a later chapter.

Staff must observe continuous improvement in the leader. The principal cannot simply repeat surface-level statements and suggestions but must continually gather evidence from the students of what good math should look like. The principal's communication, or his or her "story," should become increasingly detailed and sophisticated.

Sometimes the resistance is expressed by suggesting that the work is too difficult, that it is unreasonable to expect this of every teacher, or perhaps that the expectation that students can all rise to the desired levels is unreasonable. Clearly this underestimation of teacher capacity or student capacity is about more than math and is not helpful to the system. Focusing on schoolwide tasks and audit tasks, where change is achievable in a reasonable amount of time, will help undo this resistance. Notice the change in just a three-month period in Figure 3.2 (Duff & Tonner, in press).

Figure 3.2 | **Improvement**

Order the following 5 numbers from least to greatest. Use a model and words to support your answer.

2/5, 4/3, 5/6, 1, 1/7

A B C D E
1, 1/7, 2/5, 4/3, 5/6

I think it's that way because 1 is a small number so I put the one at the bottom and I put 5/6 at the top because they are higher than the others

Check your answer using a different model!

Order the following 5 numbers from least to greatest. Use a model and words to support your answer.

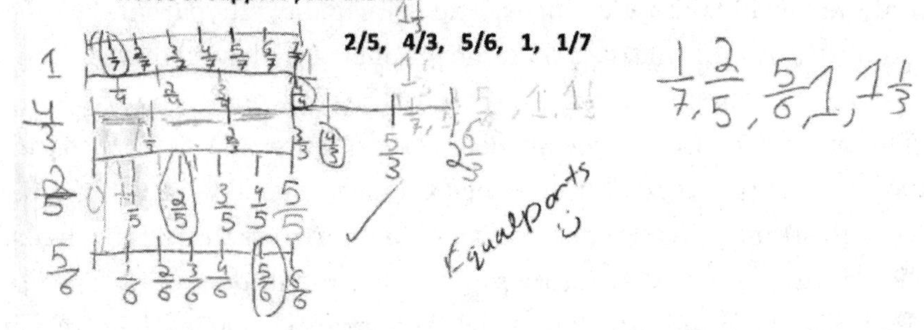

Source: From *MathUP School: School Improvement Cycle for Mathematics*, by D. Duff and J. Tonner, in press, Oakville, Ontario, Canada: Rubicon Publishing. Copyright 2018 by Rubicon. Reprinted with permission.

Getting Parent Buy-In

As discussed in the Introduction, some members of the community, including parents, have doubts about changes that have been advocated by educators in mathematics instruction. It is incumbent upon the principal of the school to both inform parents and enlist their support in increasing and improving student learning.

Although there is evidence that parental involvement is correlated with achievement (Wilder, 2014), the goal is not simply having parents visit or attend functions at the school. Instead, we want them to share high expectations for improvement in mathematics and work with the staff to effect change. Often parental involvement in mathematics involves helping children with their homework or making comments, sometimes unintentionally, about the parents' (often negative) beliefs about mathematics ("I know this is boring/hard; don't worry—it will be over soon"). The goal in improving student performance is to have the parents interact with their children more intentionally and more in line with the direction of the school.

A principal might introduce new changes at the first Meet the Teacher Night, after parents have visited the classrooms, by sharing with them the school's mathematical goals for the year and student data as the rationale behind the need for those changes. The principal elicits parental support by committing to sharing data with the parent council over the year and by requesting their support in working toward the same goal.

The principal can use social media (see Figure 3.3) to keep parents informed about progress, even sharing examples of good work. Throughout the year, the principal can repeatedly point out progress that has been made by allowing teachers to share stories and by sharing student work. The principal can also share and encourage teachers to share strategies they are using and the types of problems they are solving so parents can help. The principal can also make mathematical work visible when parents come into the school.

A principal might develop, with staff help, a summer numeracy challenge for students where students, possibly with their parents, work on achievable but interesting challenges. Students can vie for prizes in accomplishing many of these tasks over the summer, perhaps with a different task posted each week:

- **Elementary school math challenge:** Choose a number. Represent it 10 different ways. Tell how each way shows you something about the number.

Figure 3.3 | **Twitter Feed**

FD Roosevelt PS @FDRooseveltPS · Jun 2

Full math board on a Friday afternoon! #TVDSBMath

- **Middle school math challenge:** Pick a price. Show how that price could be the sale price by telling what the original price was and what the discount is. Do this in several different ways.
- **High school math challenge:** Pick a tall tree or building. Tell how you used trigonometry to find its height.

At a school we once worked with, the summer numeracy challenge began when the school team and school community council expressed concerns over the loss of literacy skills during the summer. Students were coming back in the fall with a marked decrease in their reading levels. A school team, with community support, developed a weekly, summer-long incentive program titled The Summer Passport (Duff & Tonner, in press). Booklets were handed out to the students during the last week of school with the expectation that they were to be completed over the summer and submitted during the first week back to school. Weekly online prizes were also set up to sustain momentum throughout the summer. Rather than focusing only on literacy as was the norm, because of all of the work in mathematics that year, the principal added a Principal's Numeracy Challenge section each week. The response by the community was overwhelming, and prizes had to be added to keep up with the number of entries. Teachers noticed that school readiness was higher and that reading levels were generally maintained or only showed slight decreases. Teachers were particularly surprised by the number of students who also completed the math challenges. The common story from both parents and students highlighted the number of times they would work through the math challenges together to come up with solutions. The intent was to encourage more family reading time, but the shared stories were more about the mathematics discussions. This note was included in the district newsletter written by the CEO:

> **Everyone is a learner.**
>
> On Friday, June 23, I was witness to a gym full of learners of all ages! More than 200 parents and caregivers, together with their children, were dining on a nutritious breakfast while engaged in reading and numeracy activities. Affectionately known as "Books for Breakfast," this newly developed program seeks to support parents and caregivers in how to make books and numbers come alive with their children. The program is funded by a Community Foundation grant. At the end of the event, the school staff introduced a new summer literacy and numeracy program. All students will receive a "Summer Passport," which includes summer reading and mathematics activities to be completed each week throughout the summer. Special thanks are extended to school council members, parent volunteers, and school staff for their outstanding work in supporting literacy skill and numeracy development for our students.

It is likely that a few parents will still be resistant when the principal takes these measures to involve the parent community, but it is essential that the principal, not just the teacher, takes the time to discuss these concerns with these parents. The principal should also share with them, as much as possible, educational literature that shows the value of this approach, and, most importantly, the evidence of improvement in the work of the students in their child's class.

Summing It Up

This chapter continued the discussion about how principals need to ensure that the whole school community is on board with moving forward in mathematics. There has been attention to how to overcome the principal's own discomfort with the discipline of mathematics, how to build initial support, how to work with resistant staff, and how to engage parents. The next few chapters will focus on critical monitoring that needs to be done.

4 Looking at Students for Signs of a Strong Math Culture

A vitally important part of any principal's job is to have a pulse on what is going on in the school and to ensure that each teacher is providing the best educational experience possible. By the very nature of their work, teachers tend to focus on the singular—their own classrooms. It is the principal who sees the bigger picture and makes this panoramic perspective available to everyone. While conversations with teachers about their instruction are an important piece of the picture of a school's overall math culture, direct observation of instruction, as well as a review of student work, are critical to get the full picture.

This chapter builds on the cursory review provided in Chapter 1 and digs deeper into the indicators of a strong math culture. It offers suggestions on ways to use what is observed to meet teachers where they are and help them strengthen their practice and the school's math culture.

What you might hope to see when looking at students in the math classroom falls into four main areas:

- Student interactions with one another and the teacher
- Student use of learning tools
- Student engagement with math
- Student confidence

Student Interactions with One Another and the Teacher

There are three elements of interaction that will be explored here: math talk, collaboration, and the types of questions students ask.

Math Talk

What might it sound like or look like? Gone are the days when a K–12 teacher was expected to lecture in math class, with students obediently listening and then quietly and individually completing exercises. What we seek now are vibrant, interactive classrooms where students talk about math and ask each other and the teacher lots of questions about the math (Hufferd-Ackles, Fuson, & Sherin, 2004). Physically, we want students to sit with other children most of the time or work with other students on problems on the whiteboard (Lilejedahl, 2014). We want the talking to be as much about the math content as about how to get the answer, so the given tasks should promote opportunities for that sort of conversation. We want a much greater proportion of the voice we hear to be student voice rather than teacher voice.

In a 1st grade classroom, we might expect to see students debating about whether the number 15 is more like the number 10 or more like the number 20. We are hoping to hear students talk not just about how the numerals 10 and 15 look but more about quantity. Alternately, in a 5th grade classroom, we might see students using a Two Truths and a Lie type question, deciding which of these statements is the lie:

- It might take fewer words to read a greater number than a lesser one.
- If a three-digit number has a zero in it, you never need to say more than three words to read it.
- If a number has more digits, it must be greater.

The grade 5 conversation might sound like this.

> Student A: I think the first one is true. It takes six words to say 1,245 but only two words to say 31.
>
> Student B: I don't think a number is greater just because it takes more words to say. It takes more words to say 352 than to say 1,000.
>
> Student C: Okay, I guess it is true then since it said *might*.
>
> Student C: I think the second one is true. I tried 300, 301, 302, 330, and 340, and they took either two words or three words.
>
> Student B: I agree, since all the other numbers with zeros in them that have three digits are like those ones, since the zero is either in the tens place or the units place or both.
>
> Student A: So I guess the last one must be the lie.
>
> Student B: We should probably check.
>
> Student C: I am sure it is a lie if you use decimals, since 3.2 is less than 9.

In a high school classroom, students might debate how to change an equation so that the solution is exactly half the original value, for example, changing the equation $x^4 = 100$ to $16x^4 = 100$. Their conversation might sound like this.

> Student A: I think we should use a linear equation. It's easier.
>
> Student B: Why is it easier?
>
> Student A: I think you just change the coefficient to half of what it was.
>
> Student B: I don't think so. I know that the solution to $3x + 5 = 11$ is 2, but the solution to $1.5x + 5 = 11$ is 4.
>
> Student A: Oh, maybe you double the coefficient. Let's try another one. The solution to $4x - 8 = 32$ is $x = 10$. The solution to $8x - 8 = 32$ is 5. Maybe you're right.
>
> Student C: I think we should be more impressive. Let's try $x^4 = 100$ as our original equation.

This contrasts with talk that only focuses on which procedure to follow to get a solution or talk that is only one word or a short phrase in response to a teacher's question.

Questions you might ask. As you are watching this type of talk happening, you might have opportunities to ask students questions to give you a sense of how much they have grown in their mathematical development:

- Do you like to talk about math, or would you rather just do it on paper?
- Does working on math with other students change how you do math? If so, how?
- Are the kinds of math conversations you are having now richer than the ones you used to have?

Feedback to offer or steps you can take. As a leader, you might want to encourage teachers to experiment with various strategies:

- Fostering math debates or discussions in the style of Two Truths and a Lie, which might help encourage more engaging math conversation in the classroom.
- Having students make records (on paper or digitally) of some of their math conversations early in the year and later in the year and speak to their own growth.
- Working with teachers, determine the difference between the types of tasks that generate conversation and those that don't (e.g., "Without solving, how do you know that the solution to $3x + 6 = 100$ is probably more than the solution to $10x + 30 = 110$?" as opposed to "What is the solution to $3x + 6 = 100$?").
- Changing the furniture in the classroom to facilitate conversation or purchasing whiteboards on which students can work collaboratively out of their seats.

Collaboration

What might it sound like or look like? Teachers can facilitate collaboration by ensuring that students regularly work in pairs or groups. Ideally, the groups change frequently; they are sometimes homogeneous and sometimes heterogeneous. Some teachers have enjoyed using visible random grouping (Lilejdahl, 2014), in which students are assigned to a group using an app or other strategy, such as drawing craft sticks with student names on them. This ensures that students are aware that the groupings on a particular day are determined randomly, which takes away students' concerns about why they were or were not grouped with certain other students.

It is not enough that students work in groups, however; students need to learn to be both comfortable and accountable in those situations. They need to realize that every member of the group is meant to participate fully, to be heard by the others, and to be accountable to share what the whole group finds together. It takes deliberate effort on the part of the teacher to teach those skills. The teacher's success in accomplishing this becomes visible quickly when an observer watches those students in their groups and sees honest interaction, full participation, and students listening to one another and building on their responses. For example, a teacher might tell students, "I expect each of you to do two things when you work together on this problem. Each of you must contribute to the discussion of the solution, and each of you must ask at least one question of the others in the group. After the work is complete, I can ask anyone what they contributed and what their question was." This is monitored by randomly choosing a student from the group and asking either how he or she contributed or what question he or she asked.

Questions you might ask. Consider asking students questions like these:

- Do you like having a formal role in your group, like a recorder or presenter, or do you like it to be less formal?
- Do you feel that working in a group helps you figure things out?

- Are some groups more helpful to you than others? How are they different?
- How would you choose the groups if you were the teacher?
- Do you think that all students in the group are heard?

Feedback to offer or steps you can take. As a leader, you might suggest that teachers help students by creating anchor charts with start-off questions for group work in math. These give questions students can use to begin querying other students. Start-off questions might include items like "Why did you decide . . . ?" or "What if you changed . . . ?"

You might encourage the teacher to take advice from students on how they feel collaboration in math might work better. You might also brainstorm how to ensure that some students do not dominate the conversation or how to select tasks that are appropriate for collaboration in math. If you are not comfortable enough with math to brainstorm with the teacher, turn it into a group conversation that includes strong math teachers on your staff. You can also refer to the References and Further Reading sections at the end of this book for additional resources, such as math tasks posted by YouCubed (2017) and the State Government of Victoria, Australia (2017).

The Types of Questions Asked

What might it sound like or look like? Naturally, students will ask some organizational questions, such as "Should we use pencils or markers?" or "Do you want our names on this?", but often those are the only questions students ask in a classroom. This is a signal that students are not really wondering about the content they are learning. It is a good sign, therefore, when students make more substantive statements, such as "I don't think Melissa is right because . . . " or "I think there might be other answers" or "I was wondering what would happen if. . . ."

To make this happen, teachers need to model the kinds of questions students might ask. For example, if students are working on the question "How many times do people in your family open a refrigerator door each month?", the teacher might prompt them to ask whether the month has 28 days, 29 days, 30 days, or 31 days or whether it is a winter or summer month.

Questions you might ask. Consider asking students questions like these:

- Suppose your teacher had forced you to ask him a question about the content; what might you have asked to clear up something you didn't know?
- What might you have asked to extend your learning?

Feedback to offer or steps you can take. As a leader, you might suggest to your teacher to consider using a visual to present a problem, asking each student to come up with something they wonder about when they see the visual. For example, consider the visuals in Figure 4.1 and what you wonder about when you see them.

This is a technique that is very prevalent these days (NCTM, 2016) and is a good way to get students asking questions about content.

You might suggest that teachers offer a problem or part of a problem and tell students that their job is not to solve it but to ask a question about the problem or make it into a problem. For example, students might be required to ask a question about any of these problems:

- Your brother is five years older than you. How old could you each be?
- You bought three identical items at a store. Your change was $5.12. What did each item cost?
- You want to know the surface area of a cylinder that has a height that is the same length as the radius of the base. How could the surface area be about 100 cm^2?

Figure 4.1 | **Visuals**

Visual 1:

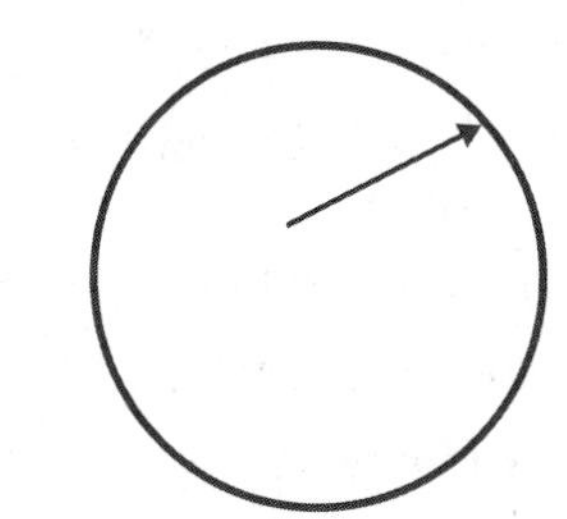

Visual 2:

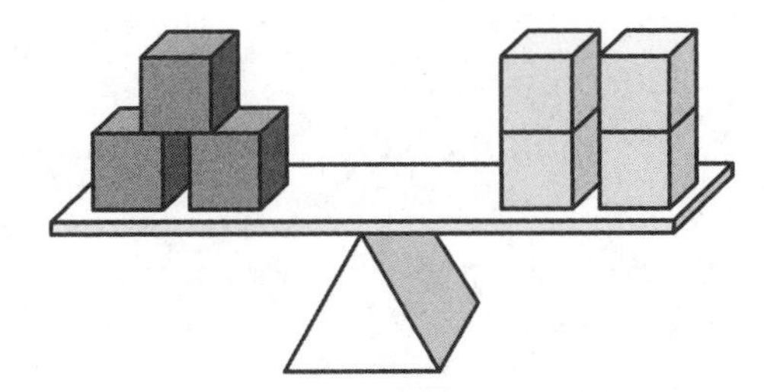

Visual 3:

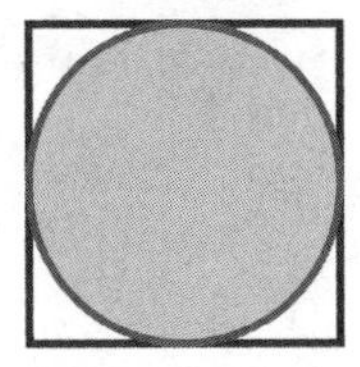

Alternately, students might turn each of these into a problem:

- You added two numbers.
- There were 250 books in the library to be divided into equal piles.
- You graphed two lines. They intersected at (1, 5).

Student Use of Learning Tools

Learning tools in math include manipulative materials, visual representations of mathematical concepts, and technology (see Figure 4.2). For many years, the mathematics education community has been encouraging the use of manipulatives as students learn mathematics, whether those students are in kindergarten or in high school (Sowell, 1989).

As technology has developed, we have added to the list of desirable materials many technological tools, such as dynamic geometry software (which allows students to perform geometric constructions and more), spreadsheets, apps, and the internet, where students can research interesting information that they can use for math.

The use of technology or manipulatives, in and of itself, does not necessarily foster a rich learning environment; it is always about how the tools are used. They should be used as thinking tools, rather than answer-getting tools.

Figure 4.2 | **Manipulatives and Tools**

PATTERN BLOCKS

LINKING CUBES

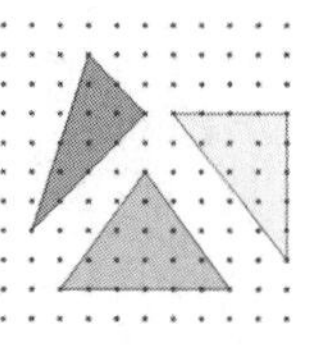

GEOBOARD

ALGEBRA TILES

BASE TEN BLOCKS

DOT PAPER

For example, calculators should be used judiciously. Students should be discouraged from using calculators for work they can do in their heads but should be encouraged to use them when the focus is not on the calculation itself but on using the calculation in pursuit of solving a problem. At lower grades (grade 5 or below), the use of calculators should be minimal.

Using Technology

Educators across the content areas struggle to determine when technology truly enhances learning and when it is merely a flashy add-on that doesn't contribute meaningfully to learning. See Figure 4.3 for examples

Figure 4.3 | **Uses of Technology**

Surface-Level Use of Technology	Richer Use
Completing digital worksheets	Watching animations of real-life events that allow students to explore proportions, such as a person figuring out how long it takes to clap 500 times if the video shows how long 30 claps take
Using calculators for work that should be performed mentally, such as 4×9 or 20×30	Using calculators to assist students to answer real inquiries that would be hard to access otherwise. An example would be, "If all the people in New York City stood shoulder to shoulder, how large of an area would they fill?"
Replacing instruction focused on conversation with instruction that just shows students what to do	Using animations to explore how shapes are built or go together
	Using virtual manipulatives to supplement physical manipulatives
	Using apps to record student explanations digitally
	Using graphing software to explore the effect of changing the values of a, b, and c in the equation $y = ax^2 + bx + c$
	Using the internet to access data to explore authentic inquiries, such as "How many planes leave O'Hare airport in a year?"
	Using download bars on a video to estimate fractions, such as "What fraction of the video has already played?" or "If 12 seconds have passed, about how long is the whole video?"

of surface-level and rich use of technology in mathematics. Teachers would benefit from having this chart as well.

Figure 4.4 provides questions educators can use to evaluate technology use in their classrooms.

Figure 4.4 | **Technology Discussion Chart**

How Are Your Students Using Technology?
Are you using technology to capture thinking?
Are students predicting before using technology?

- Before graphing, have the students predict what the graph might look like.
- Before using a calculator, have students estimate first.
- After counting the number of blue pattern blocks needed to cover a design, have students predict how many green ones will cover the same design. (Green ones happen to be half the area of blue ones.)

Using Manipulatives

Manipulatives should be used to foster deeper mathematical thinking. For example, you can use base ten blocks procedurally to model the standard addition algorithm, first modeling both numbers and then putting the same types of blocks together. While this is helpful, it keeps learning at the surface level. To use manipulatives to foster deeper learning, see the suggestions in Figure 4.5. You might share a chart like this with your teachers.

Questions you might ask. Consider asking students questions like these:

- How do you use linking cubes differently now that you're in grade 5 than when you were in lower grades?
- Which manipulatives make explorations in math easier for you? How do they do that?

Figure 4.5 | **Manipulative Use Chart**

Figure 4.2 provides images of these manipulatives.

Manipulative	Surface-Level Question	Deeper Question	The Difference
Base ten blocks	Show the number 452 with base ten blocks.	How can you show the number 456 with 42 base ten blocks, rather than with just 15 blocks (4 hundreds, 5 tens, and 6 units)? Could you use 40 blocks?	The deeper question is richer, as it forces students to use mathematical understandings. The only way to use more than 15 blocks is to trade a ten or hundred block for 10 smaller ones. (The student may need to "trade" a few times to get the required number of blocks.)
Pattern blocks	If the hexagon block has a value of one, what is the value of the trapezoid block?	If you designed pattern blocks and wanted to show halves, thirds, fourths, fifths, sixths, and tenths, what would your blocks look like?	The deeper question is richer because it forces students to think about alternate ways to show many fractions.
Algebra tiles	Knowing which tiles represent x, x^2, $-x$, $-x^2$, and -1, show $3x^2 - 2x$.	You showed an algebraic expression with four tiles. You added an expression and showed the result with six tiles. How could you show the sum using more than 10 tiles? Fewer than 10 tiles?	This second question is richer since students are forced to think about what happens when you add positive and negative amounts. For example, you can show $x + (-x)$ with zero tiles.

Feedback to offer or steps you can take. As a leader, you might support teachers in using technology and manipulatives by offering to subsidize purchases of learning tools that will be used to promote mathematical thinking and not just to get answers. You might purchase material that only a few teachers are using and have all staff solve some rich problems at a meeting using those manipulatives. It is likely that other teachers will soon see the benefit. For example, you might show how Cuisenaire rods can be used in grade 4 to solve problems with fractions or in grade 8 to solve problems involving the least common multiple (see Figure 4.6).

As a leader, you might question the overuse of calculators in the classroom. We believe that children do not develop number sense unless they can estimate and perform appropriate calculations mentally, not with calculators. We do not want children using calculators to find 3 × 4, even if we are fine with them later using a calculator for 43 × 312. To encourage development of number sense, though, we might provide more problems requiring children to estimate 43 × 312 instead of calculating it.

You might also encourage teachers to allow students to give oral or digital explanations, rather than always writing them on paper. For example, instead of forcing students, even in high school, to justify each step they

Figure 4.6 | **Cuisenaire Rod Problem**

Grade 4: Choose a value for the purple rod. Based on that value, what do you think the value of the orange rod would be?

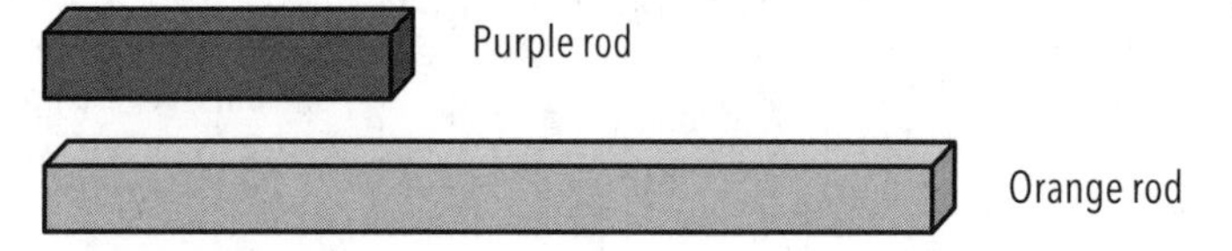

Grade 8: You create a line of purple rods and a line of black rods.
The total lengths are the same.
How many black rods could there be?

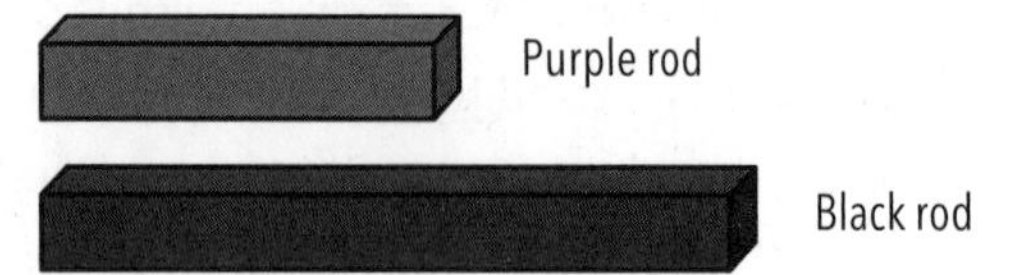

took in solving a problem, the teacher might use an app that allows the students to record their explanations. Many children are much better at explaining orally than on paper (as are many adults).

Student Engagement with Math

Often students are compliant and do what we ask of them but are not happy about the work they are asked to do. Other times, students seem enthusiastic about the tasks in which they engage, most likely because the tasks attract their interest.

Over the course of a few visits to a new teacher's classroom, a principal we once worked with observed that the teacher used many multiple-choice assignments to prepare students for the district test. Students were disengaged with the work, and, when probed by the principal, said, "I just guess or circle different ones each time." The principal had already worked on whole-school common tasks and criteria for analysis of the tasks. As a starting point for change, the principal helped the teacher repurpose the multiple-choice format to one based on the criteria they had examined in the professional learning. Instead of giving the students the questions and responses, based on the teacher's analysis of his students' work and subsequent discussions with those students of the incorrect responses from the common task, the students were given the question and were asked to develop four responses themselves. The students had to create two distractors related to common misconceptions, one correct answer, and one answer that, through thoughtful reasoning, could be recognized as not correct. Their challenge was to analyze the distractors carefully and provide reasons for their choices. Some of the students were so engaged with their work that they quizzed other classes to prove the reasonableness of their choices.

Long-term success in math is more likely when students are so interested in the math tasks assigned that they try to make sense of them and figure things out. This section will deal with student enthusiasm for math and interest in having math make sense to them.

Enthusiasm

What might it sound like or look like? Generating enthusiasm does not require a teacher to be a performer, but it does require the teacher to think about what kinds of problems will intrigue the students. A problem such as "How many text messages do you think you send in a year?" will generate more enthusiasm than a worksheet of multiplication exercises, yet both involve the same operation. A question such as "When I drink a bottle of soda, how much sugar am I ingesting?" is more interesting than solving a straightforward rate problem. Frequently, working on a problem that involves "playing" with manipulatives is intriguing, such as "If a red pattern block has a value of 3, how could you make a design with a value of 22?"

One of the advantages to teachers currently is that many interesting problems are shared on the internet and are easily accessible. One popular type of problem is called a *three-act math task,* which originated with a former high school teacher named Dan Meyer (Meyer, 2011). In Act 1, students see a video that piques their curiosity about a real-world event. In Act 2, a problem is posed, which is tackled by the students, and Act 3 brings the resolution. These are posted online for all grade levels and can be found by searching on the internet for *three-act math.*

Another popular type of problem that gets students engaged is called Which One Doesn't Belong? A variety of items are posted, and students come up with reasons why one of them might not belong with the others; there are always options. For example, for 5 × 7, 9 × 4, 4 × 7, and 124 ÷ 4, one student might decide that 5 × 7 does not belong, either because both factors are odd or because no factor of 4 is involved. Another student, however, might decide 4 × 7 does not belong, because the result is less than 30. A third student might decide 124 ÷ 4 does not belong because the operation is division and not multiplication, and a fourth might decide that 9 × 4 does not belong because it is the only one with a result that is a perfect square. The sets under consideration can contain shapes, numbers, computations, graphs, or other items. Teachers can search the internet using the phrase *Which one doesn't belong?* and find many possibilities.

Questions you might ask. Consider asking students questions like these:

- What is the most interesting problem you've done in math recently? What made it interesting to you? How did it make you think differently?
- Do you usually look forward to math activities? Why or why not?

Feedback to offer or steps you can take. As a leader, when you see a lack of enthusiasm among students, you might ask teachers where they are finding the problems they are assigning. Perhaps you can suggest a three-act math task or a Which One Doesn't Belong? problem. You might also take advantage of social media and send a teacher or group of students an interesting problem, puzzle, or activity that fits with what you observed in the class. You might suggest, "I thought this would be perfect for your class. When you're working on it, let me know so I can come by or send a student down when they have an interesting solution." Those solutions might become the springboard to a hallway conversation, an introduction to a staff meeting, or a reply to a parent.

Making Sense of Math

Once you have students more engaged, it is easier for them to rise to the challenge of deep understanding and making sense of math. Ideally, the goal of most math activities should be making sense of mathematical ideas. In fact, the first Common Core Standard for Mathematical Practice (Council of Chief State School Officers & National Governors Association, 2010) focuses on the notion of asking students whether solutions make sense.

Students should want to understand what is going on. You do not want to hear students say, "Just tell me what to do"; you want them to *want* to figure out what to do.

What might it sound like or look like? You should expect that, when students are asked to explain their thinking, they will not see this as an unusual or unnecessary burden. This happens when teachers simply make this practice a habit and learn to prompt students appropriately.

For example, suppose students are asked to determine the original price of an item that now costs $98 after a discount of 30 percent. To help teachers create prompting questions, as a principal, you might connect back to models or ideas you've been working with across the school. You might prompt for a number line representation or the use of a benchmark as the students move through the estimate. You might focus students on relationships and ask what percent is paid and what other percentages they could figure out from that information. Students realize that 70 percent is paid and that is $98. They might benchmark to one-third because half of 70 percent is easy to determine and half, or 35 percent, is close to one-third. So you might hear them think aloud: "That means that 35 percent is $49 and that is about one third. So the full price must be about $50 × 3 = $150. I'll start there."

Sometimes students learn by giving unreasonable answers, telling why those are unreasonable, and telling how this helps them know if their chosen answer seems reasonable. By connecting the thinking back to common models, teachers are continually reinforcing mathematical thinking, rather than only solving the problem.

The teacher might create prompting questions like these:

- Do you think the original price was more than $100? Why do you think that?
- Do you think the original price was close to $200? Why do you think that?
- What percent of the original price did you actually pay?

What an observer would see in this kind of classroom is that explanations seem natural, not a list of steps a student ticks off while going through them.

Suppose students are asked why 23×45 is not equivalent to $20 \times 40 + 3 \times 5$. We want the students to create a picture or use words to show why or give some other reason, such as "23 groups of 45 is the same as 20 groups of 45 and 3 groups of 45, not 20 groups of 40 and 3 groups of 5." Alternately, students might be asked to compare how they would solve the equations $2x + 3 = 45$ and $2x + 3 = 3x - 4$. We want students to realize that, in each case, the goal is to create a simpler equation that must have the same solution. For example, $2x + 3 = 45$ has the same solution as $2x + 4 = 46$, but the second equation is no simpler. On the other hand, $2x = 42$ really is simpler.

Questions you might ask. Consider asking students questions like these:

- Do you think most math ideas are explainable? Why or why not?
- If I asked you to explain why your answer makes sense without repeating your steps, what would you say?

Feedback to offer or steps you can take. As a leader, you might note whether student explanations are mostly just a recap of steps they took and therefore not really explanations. You might find professional development support for teachers who struggle with explanations themselves, whether by reading a print resource such as Marian Small's *Making Math Meaningful to Canadian Students, K–8* (2017b), by working with a colleague, or by completing a professional development session.

As a leader, you might suggest a list of questions teachers could use:

- Can you use an example to show what you mean?
- How does that model support your solution?
- Here's an alternate argument . . .
- Ali disagrees with your solution. How do you respond?

- How does your answer connect to Ethan's?
- How did you use different models and find a similar answer?
- How are all these problems similar?

Student Confidence

Our goal in education is to provide students with the tools they need to eventually "stand alone." Of course, we support students along the way, but our ultimate goal is to have faith that they can figure things out for themselves. That means they need many experiences working through mathematical situations that are appropriate at different levels, and it also means that they visibly see their teachers having confidence in their ability to succeed.

When a student asks for help, the teacher should not do the work for the student. Scaffolding needs to be at an appropriate level, not giving away too much (Dale & Scherrer, 2015). The teacher needs skills in asking the right questions to move the student forward. In different situations, they might ask, "What do you think the problem means?" or "What kind of answer do you expect?" Some of that skill comes from knowing the learning progressions for various concepts. Many math educators have attempted to create learning progressions, and some are particularly accessible, such as Graham Fletcher's *The Progression of Addition and Subtraction* (2016). The Common Core Standards Writing Team of the University of Arizona Institute of Mathematics and Education has also prepared detailed progressions for the Common Core standards (2013). Progressions may also help with the next question to ask, whether to challenge or to scaffold. A principal should observe whether the focus is on teaching students what they already know or finding out what they don't know and attending to the next phase of learning.

Confidence will be examined in three aspects: student independence, student perseverance, and student creativity.

Independence

We do not want students who constantly seek confirmation that they are doing what the teacher wants; we want students who feel confident enough to make some of their own decisions, knowing that their teachers are comfortable with that. The principal will recognize that, if the teacher lectures too much, the students will generally seek confirmation because all the knowledge is transmitted from the teacher and little accountable talk comes from the student. The onus on proving and reasoning should be on the student, with the teacher simply pushing for clarification and justification or proposing challenges to the argument. ("Clara is saying. . . . How does that fit into your argument?")

Questions you might ask. Consider asking students questions like these:

- I noticed that you just asked your teacher whether you were right or not. Were you fairly sure before you asked?
- When you are not sure what a problem means, do you usually ask your teacher first, ask a friend, or try to figure it out yourself?

Feedback to offer or steps you can take. As a leader, you might suggest teachers adopt a policy where they don't instantly respond but turn questions back to the students, as in "What do you think?"

Perseverance

We want students who have enough confidence that they will not quit the minute they run into difficulty. Students should expect to run into obstacles and expect to try again with a new plan. Perseverance is an important academic asset (Duckworth, Peterson, Matthews, & Kelly, 2007). Perseverance is built on experiences where perseverance pays off.

That means teachers need to provide problems in each child's zone of proximal development, or the area in which students do not already know the answer to a problem but are close enough to be successful working on it (Vygotsky, 1978). This may mean different problems for different students in the classroom.

What might it sound like or look like? In too many classrooms, the same problem is assigned to all students. For some, however, it is simply too far beyond their cognitive level, and perseverance will not pay off. It is important that teachers consider this and rework problems as appropriate. One way to accomplish this is to create parallel tasks and allow students to choose the one with which they are the most comfortable. For example, consider this task:

> Choose one:
>
> 1. A linear function
> 2. A quadratic function
> 3. An exponential function
>
> Work with a partner to describe each of these, using your chosen function: the values of your function for an input of $3x$ and for an input of $x + 3$, as well as 3 more than the value of your function for an input of x and 3 times the value of your function for an input of x, i.e., $f(3x)$, $f(x + 3)$, $f(x) + 3$, $3f(x)$. Tell how you know you're right. How are your answers similar? How are they different?

More examples can be found in Small and Lin (2010) or Small (2017a).

Questions you might ask. Consider asking students questions like these:

- How long do you usually work on a problem before you give up? Does it depend on the problem? How?
- What is it about that problem that made you give up?

- Are there some kinds of problems you give up on more quickly? What kind?
- What part of the problem made you give up? What parts did you feel comfortable with?

Submitting a blank page for a common task is a form of giving up. It is also a sign of math culture in the room. We want the culture to be "It's OK not to know, but it's not OK to give up." You can tell a culture of the room or school by the number of blank pages you see when looking at student work.

Feedback to offer or steps you can take. As a leader, you might, based on your conversations with students, help teachers become more aware of the aspects of a given problem that led various students to quit. You might also note if a specific problem seemed to be clearly too far beyond a particular student. You might ask the teacher how to adapt the problem to make perseverance pay off.

Creativity

We want students who like to put their own "spin" on the work that they do. Many adults, because of the way they learned math, do not see math as a place for creativity; in fact, students should be provided with opportunities to be creative in math (Silver, 1997). Again, this is built on student confidence that the teacher appreciates creativity and is open to looking at things in unique and personal ways. It is also built on experiences where teachers name the strategy an individual student uses to show appreciation for it.

What might it sound like or look like? You might see a situation where a student is given a clearly creative task, such as making a video that would help another student understand the concept of prime numbers. However, you might also see a situation where a student simply has a unique solution strategy or makes unique observations. For example, a 3rd grader who is asked how the numbers 350 and 550 are alike writes, "Neither are degrees of latitude or longitude." Hopefully, the teacher appreciates this creative response.

Questions you might ask. Consider asking students questions like these:

- Do you feel there is an opportunity to be creative in math?
- Do you feel that students who are successful in math are more creative or less creative?
- What is the last time you gave a creative response to a math problem?

Feedback to offer or steps you can take. As a leader, you might discuss in staff meetings the importance of not making a distinction between "math people" and "creative people" and encourage all teachers to set tasks that encourage creativity.

Summing It Up

This chapter detailed what the math culture in a school should look like when observing students during instruction. There has been attention to student interactions with each other and with the teacher, student use of learning tools, student engagement with the math, and student confidence. These are detailed in Appendix A. The chapter also offered suggestions for questions you might ask of students, feedback you might offer teachers, and steps you might take to support teachers in improving these facets of instruction.

5

Looking at Teachers for Signs of a Strong Math Culture in the Classroom

This chapter explores what a principal should expect to see when observing a teacher in an instructional situation and offers suggestions on the kinds of responses the principal might make. Principal observations of teachers should focus on five main areas:

- Thoughtful use of learning time.
- Appropriate lesson styles.
- Fostering deeper understanding.
- Teacher confidence in students.
- Appropriate and timely learning support for students.

Thoughtful Use of Learning Time

Because instructional time in math is limited to a certain number of minutes per day, that time needs to be used efficiently to maximize student learning. Many teachers struggle to get through the entire curriculum for which they are responsible, and often time still runs out. This can happen for a variety of reasons:

- Classroom management routines that do not create a good pace for lessons.
- Long periods of in-class homework review time, even when students did not struggle with it.
- Poorly written problems that take too long to complete.
- Too many students sharing their work, even when the relevant points have been raised by the first few responders.
- Teachers assigning a very complex problem and waiting for too many students to complete it fully, rather than taking the time first to diagnose whether students are ready for such a problem.
- Lessons that get derailed by side issues that aren't related to the learning goal.

Good classroom management techniques (see Figure 5.1) are relevant to teachers of all subjects, not just math. There is evidence that student achievement improves when rules and procedures are clear, when disciplinary interventions are minimized, when disruptions are handled appropriately by teachers, and when the teacher is always mentally ready to deal with management issues (Marzano, Marzano, & Pickering, 2003).

Figure 5.1 | **Wasted Time**

To avoid wasted time, do you

- Have strategies for responding quickly to attention-getting moves?
- Have materials readily available?
- Have strategies to spend minimal time reviewing homework?
- Make sure problems are clearly posed?
- Ensure, in advance, that students are ready for your task?
- Avoid getting derailed by side concerns?

In many math classrooms, particularly in higher grades, teachers spend the first part of the period, often up to 20 percent of their instructional time, reviewing homework, usually problem by problem (Otten, Cirillo, & Herbel-Eisenmann, 2015). Traditional review of homework focuses on answers rather than the big mathematical ideas students need to learn or on student misconceptions. Otten and colleagues (2015) recommend "talking across problems," that is, talking about issues that arise in multiple problems and talking about errors rather than correct solutions; this allows the teacher to focus on the intended ideas behind the problems. Consider this sample problem: *The cosine of an angle is about 0.31 less than the sine. What could the angle be?* Instead of asking for the correct answers, the teacher might ask questions like "What did you notice about the angles when the sine was greater?", "When was the cosine greater?", or "Could the sine and cosine ever be really close? When?"

Many of us have been in a math classroom where a problem is posed and nobody really understands what is being asked; sometimes even the teacher has difficulty articulating the task for the students. There is evidence that many teachers struggle with posing good problems (Chapman, 2012). When a teacher poses a poor problem and the group wastes its time just figuring out the problem itself, learning time is lost. Poorly written problems have some common characteristics:

- Unintended lack of clarity that could be easily fixed
  A pattern block design is $\frac{2}{3}$ yellow. How many blocks are there? The student does not know if $\frac{2}{3}$ of the blocks are yellow or if $\frac{2}{3}$ of the area is yellow. The student does not know if the total number of blocks is required or only the number of yellow blocks.
- Lack of clarity that is problematic and not easily fixed
  What are you supposed to do when you add numbers? The student does not know what kinds of numbers are being referenced, whether a procedure is being asked for, or what is really expected.

- Difficult vocabulary
  The dividend has three digits, and the divisor has two digits. What do you know about the quotient? Some students see those words and just turn off. Even though a teacher realizes it is important to learn correct vocabulary, the question is when the words should be used; if used too early, they become a barrier.

With good intentions, generally in the interest of showing the students that each of them matters, many teachers ask many, if not all, students to share their thinking. Although it is critical that all students have the opportunity to be heard, it is certainly not essential and not a good use of time for each student to be heard for each solution. Teachers must thoughtfully sequence and orchestrate sharing to maximize the effectiveness of the discussion (Smith & Stein, 2011). Teachers must think about which strategies to share and in which order, often beginning with a more familiar one and moving to more sophisticated ones. Also, teachers need to consider carefully which students should present on behalf of a group. In choosing the students to present, the teacher needs to balance the need for all students to be accountable with the recognition that some students may struggle to communicate effectively to the whole class.

No matter what task is posed, the teacher must decide how long to let students work before there is either the expectation of an answer or discussion of an approach to the problem. In a problem-based teaching environment, a substantial problem often has students working for 20–30 minutes; however, the teacher must decide how much time is enough. If the focus of the problem is on the mathematical thinking more than it is on the execution of the final steps toward a solution, then perhaps time is better spent stopping earlier and focusing on the thinking more, addressing the final solution later. As a side benefit, this approach allows all students, even those who have not completed the problem, to participate in the discussion of the problem.

Most important, keeping the learning goal front and center throughout the lesson will help teachers use time wisely. Sometimes that will mean disclosing the goal to students early in the lesson so that the students know what is coming. Other times, though, the goal really cannot be disclosed until the end, or it might give away something students must figure out on their own. Whether the goal is disclosed to the students or not, the teacher must remain "on goal" all the time.

As you are watching a lesson unfold, you might ask these questions:

- How much time do you spend on going over homework? Is that worth the time it takes?
- What are the key points you are planning to uncover in the homework discussion?
- I notice that a lot of students were not really listening to solutions after the first one. Have you considered alternate ways of sharing?
- I noticed that a lot of students were waiting a long time before you discussed the problem. Have you considered strategies that might get to the sharing earlier?

As a leader, you might encourage teachers to choose an area of focus to help them use time more wisely, whether it is sharing solutions, management issues, taking up homework, or some other issue. Have teachers monitor their own improvements in that area over a one- or two-month period. You might lead a discussion with teachers about the types of homework they are assigning and encourage teachers to share alternate ways to check it (or even if it needs to be checked). You might refer teachers to the article by Otten and colleagues (2015).

Lesson Styles

It is ideal if teachers frequently teach through problem-solving strategies and pose problems that either pique curiosity or lead to generalizations and transfer. It is also important that teachers focus on building understanding, not just the repetition of procedures, and ask questions to build students' ability to think, not just mimic what they are shown. This section will discuss several important practices:

- Lessons built on problem solving
- The types of problems teachers use and how interesting they are
- The use of open-ended tasks

Teaching Through Problem Solving

What might it sound like or look like? To begin creating a lesson, a teacher needs to establish a learning goal, rooted not only in the curriculum but also in important mathematical ideas. The teacher then locates or develops a problem that would lead to that learning goal and carefully considers how to emphasize that goal while discussing the problem. The teacher might then go back and create an activation that leads nicely into the selected problem. To determine these learning goals, teachers need to think about what the curriculum requires and about appropriate sequencing. The problems come after the fundamental planning has been completed. There are four steps to creating a rich math lesson:

1. Start with a learning goal rooted in curriculum as well as important mathematical ideas.
2. Locate or create an appropriate problem to lead there.
3. Develop questions about the students' work on the problem that brings out the learning goal.
4. Go back and create a short activation activity to lead nicely into the problem.

Teaching through problem solving involves posing an appropriate problem to bring out the ideas students need to explore by letting them figure the problem out, rather than showing them how to solve it first. For example, rather than showing students how to multiply two 2-digit numbers (assuming they already know how to multiply by 10 and by 1-digit numbers), the teacher poses a problem that is solved by multiplying two 2-digit numbers and lets students come up with their own strategies. It does not mean that teachers are sitting on the sidelines; teachers will be probing, scaffolding, and helping with connections, but students will do the heavy lifting. Students learn best in this kind of classroom (Boaler & Staples, 2008).

A principal might watch for attention to differentiating instruction and for questions to relate that day's ideas to other ideas previously learned or ideas that are coming. A principal might also watch for probing questions that help everyone understand better what students are trying to communicate. Ideally, these probing questions are rooted in a teacher's knowledge of individual students and their previous work; the teacher should be able to articulate the connection between those questions and the individual student's needs and performance.

Questions you might ask. After an observation of a problem-based lesson, you might ask:

- What did you do to make that problem more accessible to ___, who usually struggles?
- How often are you teaching through problem solving? How do you make sure you are stretching all students?
- How has student work changed since you've been teaching this way?
- Are there misconceptions from previously gathered data that you will be looking for?

Feedback to offer or steps you can take. As a leader, discuss with staff the need for the problems they select to be appropriate for the students' level, interesting, and relevant to the curriculum. You also must continue to encourage them to *probe,* rather than *tell,* as students work on problems. You might ask how the problems chosen ensure that students learn new ideas (not just assess what they already know) or how the problems fit into a sequence of learning.

Posing Problems That Pique Curiosity or Lead to Generalizations

Teachers always have choices in the selection of problems for their students, but it is important that these problems either pique curiosity or lead to bigger mathematical ideas (Piggott, 2008).

What might it sound like or look like? For example, students trying to decide how much it costs to raise a child are engaged in a much more significant mathematical experience than students determining the change for a particular purchase and are usually much more curious about the result. Other problems that might pique curiosity at different levels might include the following:

- How many blocks are usually in a tall tower?
- How long is hair, in inches, before you say it's long?
- About how many apples trees would you need to plant to get 1,000 apples in a season?
- About how big of an area would you need to fit all the people of the United States?

Problems that lead to generalizations are also important to include in a rich math program. For example, solving a problem such as "The perimeter of a rectangle is triple its length. What could the dimensions be?" leads to generalizations, because there is a pattern to these rectangles. (The length must be double the width.) Students are likely to explore several possibilities and then look for patterns and relationships. Generalizations occur either through examples that students create (such as 5 × 10, 4 × 8, and 10 × 20 rectangles) or by reasoning even more abstractly (noticing that if the length is double the width, the perimeter is the sum of the length at the top, a length at the bottom, and two half-lengths from the sides).

Here are a few other examples:

- You add two numbers, and the result is double their difference. What could the numbers be? Why do those pairs of numbers work? (One number is always three times the other.)
- You multiply two numbers that are two apart on the number line and compare the result to the square of the number between them. What happens? Why? (The product is always one less than the square of the number between them.)
- You graph a function. You interchange the variables x and y and graph the new function. What do you notice? (The new graph is reflected across the line $y = x$, although it may no longer represent a function.)

It is also worth noting whether the teacher is using open questions. Open questions allow for richer discussions, because open questions can be approached from many perspectives, an important aspect of building critical thinking. For example, the teacher might ask, "The difference of two numbers is just a little less than what you subtracted. What could the two numbers be?" This not only leads to a generalization (the subtracted number must be slightly more than half of the number you are subtracting from), but it also allows for differentiation. Some students can use simple numbers and work out a few possibilities, but other students could move directly to the generalization (see Figure 5.2).

On the surface, these problems might just seem interesting but not necessarily important. However, teachers who use problems like this are demonstrating that they have the knowledge and confidence to deal with a range of student thinking, something that is critical in an environment where students do the bulk of the work. When these types of problems are not evident in a

Figure 5.2 | **Creating Open Questions/How They Generalize**

Regular Question	Open Version	Possible Generalization
What is 52 + 28?	You add a number to about half of it. What could the sum be?	When you add a number to about half of it, the sum is always about three times the smaller number.
What is $\frac{3}{8} \div \frac{3}{5}$?	You divide two fractions with the same numerator. What could the quotient be?	When you divide two fractions with the same numerator, the quotient is always the result of dividing the second denominator by the first denominator.
What is sin 10°? What is tan 10°?	The sine of an angle is pretty close to its tangent. What could it be?	When the "opposite" side of a triangle is small, the adjacent side and hypotenuse are similar in size and sine and tangent are close in size.

classroom, it could be a signal that the teacher is uncomfortable with questions that have ranges of answers, which is something that needs to be addressed.

Questions you might ask. You might observe the types of problems teachers are choosing for their math lessons and ask questions such as these:

- Do you think there might be value in including questions that force students to find out information that is not given to them directly?
- How often are you asking open questions in math? Has it changed any student's performance?

Feedback to offer or steps you can take. As a leader, you might ask teachers to track how often they ask open questions. As a staff, you might make it a goal to increase the percentage of open questions asked in math class. If open questions are used as focus tasks, teachers can see student growth (or lack of it) over the grades. Similarly, teachers might be asked to track how frequently they base a lesson on a rich problem and the expectations they have for student modeling/communication of their work.

Fostering Deeper Understanding

Facility with solving equations or with multiplying large numbers certainly is not a priority with the growing number of technological tools that are available. What is a priority, though, is that students move beyond rote knowledge to acquire a deep understanding that equips them with strategies for approaching new situations and problems. They also need to spend time thinking about whether an answer is reasonable and how to work collaboratively with others; this is the life they are moving toward (Gomes, 2016). This section will focus on the difference between *doing* math and *thinking* math.

What might it sound like or look like? Contrast the questions in each of the pairs in Figure 5.3. The first question involves recall, but the second requires understanding.

Figure 5.3 | **Recall Questions Versus Understanding Questions**

Primary Grades:

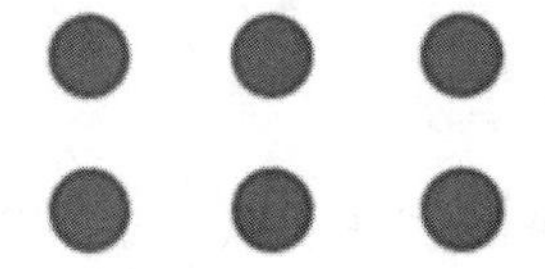

Recall: How much is this?

Understanding: How would you arrange six counters so it is easy to see that it is the number 6? How would you arrange them to make it less easy?

Grades 3–5:

Recall: Read this number: 1,002.

Understanding: You read a number and you say exactly four words. Describe a large number it might be. Describe a smaller number it might be.

OR

Recall: What is 2003 – 1249?

Understanding: How do you know that 2003 – 1249 must be around 750 without actually subtracting?

Grades 6–8:

Recall: What is 30% of 45?

Understanding: Suppose x is 80% of y. What percent of $2y$ is x?

OR

Recall: 43 is 60% of a number. What is that number?

Understanding: 43 is 60% of a number. Without finding the answer, tell whether 80 or 90 is closer to the exact value of the number and why.

Grades 9–12:

Recall: Solve: $3x^2 - 10x + 4 = 0$

Understanding: Why does it makes sense that $3x^2 - 5x + 4 = 0$ has no solutions?

OR

Recall: Solve for x: $5^{20} = 625^x$

Understanding: Which is easier to estimate: 5^{20} or $4^{\frac{2}{3}}$? Why?

Teacher assessment of learning should focus as much on understanding as the instruction does.

Questions you might ask. After observing a lesson, you might ask questions like these:

- Do you feel your focus in instruction is more on understanding or on knowledge or equally on both?
- How do you make sure you ask questions that highlight understanding? Are there certain styles you use?
- Do your assessments focus more on knowledge or understanding? Show me what you mean.

Feedback to offer or steps you can take. As a leader, you might share a table like Figure 5.4 to help teachers see the difference between *doing* questions and *understanding* questions.

Figure 5.4 | **Doing and Understanding Questions**

Level	Doing	Understanding
K–5	You gave the clerk \$10 for an item that cost \$4.58. What is your change? Which is greater: 34 or 72?	You paid for an item and got one bill and six coins in change. What could the price have been? Which is probably greater: □2 or 7□? Why?
6–8	What is $-8 - (-4)$? Find 40% of 88	You subtracted two numbers, and the difference was -4. What do you know about those numbers? What signs could they have? How large or small could they be? Number *A* is 40% of Number *B*. What number is Number *A* 30% of?
9–12	Graph $y = 3x^2 - 2x + 8$ Write 40° in radian form.	When is the graph of $y = 3x^2 - 2x + 8$ almost indistinguishable from the graph of $y = 4x^2 - 2x + 8$? Why is that? Why might it be useful to use radians rather than degree measures? When would it not be useful?

You might devote a professional learning session to a discussion of strategies like the following for developing questions for understanding:

- Ask for estimations rather than calculations.
- Ask for explanations that do not refer to the answer or solutions, such as these examples:
  - How could you estimate the solution to $3x - 4 = 2x + 10$ *without actually solving it*?
  - *Without actually multiplying*, how do you know the product of 342×152 must be around 50,000?
  - *Without referring to the answer or a "rule,"* explain why 43 + 57 must be even.
- Ask *why*.

Confidence in Students

Teaching through problem solving is a way to show confidence in students; that is, by not showing students what to do but allowing them to figure it out, teachers demonstrate confidence in their students' abilities. Teachers also show confidence in students by appreciating divergent thinking. Encouraging divergence is a way of telling students that their job is not to guess what an adult wants to hear; instead, the adult is actually curious about what the student has to say.

What might it sound like or look like? In math, confidence in students might emerge from open-ended questions like these:

- A particular multiplication involving big numbers was really easy to do in your head. What multiplication might it have been, and why was it easy?
- Some people think that it's easier to work with fractions if you change them to percentages. What do you think? Why?
- How would you create a fraction that is just *slightly* less than $\frac{2}{3}$?

Questions you might ask. As you watch the teacher interact with students, you might ask questions such as these:

- Do you feel that your students are comfortable giving unusual responses? How might you encourage it more?
- Why did you choose to use a modeled approach for that lesson?

Feedback to offer or steps you can take. As a leader, you might have teachers discuss their feelings about the questionnaire in Figure 5.5. It is important that the diverse viewpoints of the staff be acknowledged and that healthy discussion ensues.

For many leaders, the work of changing teachers' preconceived beliefs about student ability isn't as simple as listing "growth mindset" as a

Figure 5.5 | **Student Confidence Questionnaire**

Circle your response to each statement.			
I believe that students can normally figure out the math themselves.			
Strongly agree	Agree	Disagree	Strongly disagree
I believe that students will be more successful if I model for them first.			
Strongly agree	Agree	Disagree	Strongly disagree
I think that most math questions do not really allow for divergent responses.			
Strongly agree	Agree	Disagree	Strongly disagree

school goal. Many principals we know have indicated that staff members were much more receptive to changing literacy beliefs than those in math. When principals asked teachers specifically about their struggling readers, the teachers would outline detailed plans for improvement, or at least they would talk technically about the interventions in place for particular students to increase their reading levels and help them become more proficient. The teachers believed that, as a result of their teaching, students would show quantifiable progress. Interestingly, when principals asked similar questions about students who had difficulty in math, the teachers pointed to the students' inherent weaknesses and not to their own ability to change student understanding. Often there weren't quantifiable targets for improvement, as there were in literacy. In many cases, it appeared that the staff believed they had strong impact on changing literacy proficiency, but math proficiency was affected by a student's inherent ability and not by their instruction.

Changing this mindset became the work of the leader. Staff needed to see that their teaching in math could have a positive impact on student achievement. As the staff focused more on student work with defined criteria, they began to see evidence of the effectiveness of their teaching on student understanding. Principals would then highlight growth both individually and by class in the student data. During a monitoring visit, teachers demonstrated the change in student performance by showing task samples to the principal. Conversely, the principal commented to the teacher and to the individual student, on a visit or during a hallway conversation, about the positive change in the work as a result of the new learning.

Learning Support for Students

Teachers support students in many important ways—socially, emotionally, and academically. In this book, we have chosen to focus mostly on academic support because many resources are available to principals on helping students socially and emotionally, such as works by Clarkson, Bishop, and Seah (2010) and Furner and Berman (2005). However, there is not much on helping students academically, specifically in the area of math.

Teachers provide academic support in several different ways:

- Listening to students to determine their needs.
- Scaffolding strugglers rather than telling them the answers.
- Encouraging strong students to go beyond the bare minimum.
- Providing thoughtful feedback to all students.
- Highlighting the important math for all students.

Listening

One way for a teacher to support students is to listen to them. Often teachers get so focused on getting students to where they want them to go that they don't take the time to listen and appreciate fully where the student is on the path to mastery. Teachers often assume that students are interpreting what they say the way they intended, but that is not always the case. Listening and asking probing or clarifying questions is essential to using what students say as a vehicle for assessing their knowledge.

What might it sound like or look like? This is an example of a conversation that might have occurred in an actual classroom where a teacher intended the student to talk about quantity, but the student assumed it was about numerals, and neither knew what the other was thinking.

> Teacher: Which is bigger, 55 or 85?
> Student: 55.
> Teacher: Why did you say 55?
> Student: Because it is bigger.
> Teacher: But how do you know it's bigger?
> Student: Because it looks bigger.
> Teacher: What do you mean by "looks bigger"?
> Student: It is hard to see the 85, but easy to see the 55 over the door.

By taking the time to listen and not jumping to conclusions, the teacher discovered that the young student was talking about numerals he had seen on houses in his neighborhood, and the numeral 55 happened to be bigger.

Questions you might ask. Questions like these can be used to discuss a teacher's listening practices:

- I noticed you spent a lot of time listening to several of your students. How important is that to you? Why?
- Have you thought about ways you might engage other students so you have more time to talk to individual students?

Feedback to offer or steps you can take. As a leader, you might encourage each teacher to choose at least one student per day and plan on spending at least three to five minutes of dedicated time listening to that student. The student could provide a response to a math problem/question or simply be talking while working on a problem.

As part of her work on small-group math interventions, one principal we worked with encouraged staff to establish groups of two to four students who had common strengths or misconceptions demonstrated by their task data. The teacher took the first part of every group session to make a video recording of one student's reasoning before addressing the focused teaching. The group would then listen to the reasoning. The teacher sometimes offered scaffolded prompts for the students to follow:

- Explain your reasoning.
- Which models did you use to show your thinking?
- How did you check for reasonableness?
- How did your thinking connect with what you have learned in the classroom?
- How does your oral explanation compare to your written response?

Students in the group could take a moment to consider the explanation, add to the thinking, or connect the ideas to their work. Following this routine, the teacher would highlight the thinking in a short presentation to deepen the group's understanding of a particular concept or to address a misconception. Often, parts of the video were used by the principal at a staff meeting or professional learning session to show common misconceptions or to elicit other responses on how to connect the key idea across grades.

Probing and Scaffolding Rather Than Telling

When students struggle, teachers can support students by refraining from telling them what to do and asking appropriate scaffolding or probing questions instead. Probing questions are questions that have students restate, elaborate, or clarify their ideas. Kosko (2016) indicates that teachers who focus on understanding are likely to ask more probing questions than other teachers. Sometimes probing questions are generic, such as "Can you say that another way?" or "Tell us how you got started; why did you start that way?", but sometimes probing questions are specific. For example, suppose a student is asked to determine two numbers that are 40 apart on the number line and add to 150. The student says that the numbers are 55 and 95. The teacher might probe more specifically by asking, "How did you know that both numbers had to be more than 50?" or "How did you know both numbers were less than 100?"

Scaffolding is used to assist a student who is unable to reach a solution independently; it is particularly important that scaffolding be intentional and sensitive to student needs. There is evidence that even teachers who use rich tasks often transform them into less challenging tasks during instruction by over-scaffolding and breaking the rich task down into a multitude of smaller tight tasks (Stigler & Hiebert, 2004). With good scaffolding, the student is still doing the hard work in figuring out the problem. The more teachers do the thinking for students, the less students will learn to think for themselves. When students are struggling with reading, we discourage sounding out every letter or word for them; the same should be true in math, with students being encouraged to meet the challenge.

Suppose, for example, a student is struggling to figure out how long it would take someone to drive 512 miles if the person drives 40 mph for the first two hours and then 60 mph for the rest of the trip. A student may want to average 40 mph and 60 mph and divide 512 by 50, which does not give the correct answer. Instead of showing the student how to solve the problem, the teacher might ask questions like these:

- Do you think he was halfway into his trip when he switched to 60 mph? Why or why not?
- Do you think he could have driven for 12 hours? Why or why not?

Scaffolding, just like probing, should emphasize questions and not simply model procedures or tell a student too much.

Questions you might ask. Consider asking teachers questions like these:

- How have you been avoiding telling students too much when you scaffold?
- How long do you wait before you scaffold? Do you wait the same amount for all students?
- Do you have more success using generic probing questions or ones specific to the task?

Feedback to offer or steps you can take. As a leader, host a conversation where teachers consider samples of student work and discuss how they might probe or scaffold that work. Challenge teachers to ensure that their scaffolding supports students but does not overly direct.

As many school principals and staffs that we've worked with learned more about the importance of proving and reasoning, they included a routine of scaffolded questions that helped students think about some sort of generalization. Principals began the work informally within a class visit when an individual student presented his or her reasoning to a problem. The principal used four to five probing questions for the student to consider while working through the explanation.

- Why do you think that your answer/explanation is reasonable?
- Does that work every time?
- What if we tried . . . ?
- Can you find a case where it doesn't work or one you're not sure of?
- Why do you think this reasoning works every time? What can we say about all these cases?

The principals communicated some of the work in the staff bulletin, school newsletter, and sometimes on the announcements. What they noted was that teachers began using the routine as part of their student conversations and began to change *doing* questions into thinking or reasoning questions to uncover relationships and promote student generalizations.

Providing Extensions for Strong Students

Strong students also need support. Rather than simply having them read a book, help another student, or solve the problem another way when they finish quickly, teachers need to provide them with opportunities for challenge. Appropriate extensions or alternate challenges should be available to them. For example, if students had been creating scale drawings and a student finished his or her drawing quickly, the teacher might, either in place of or in addition to the original question, ask questions like "If you were drawing a map of Illinois on a piece of loose leaf paper, what would an appropriate scale be?" or "How would the scale change if it were a map of Alaska?"

Often problems that lead to patterns and generalizations are particularly interesting to strong students:

- You add two numbers and the answer is triple their difference. What could those numbers be?
- You are creating a shape where the perimeter is four times one of the lengths. What kind of shape is possible?
- A price reduced by 20% is equal to a different price reduced by 40%. How were the original two prices related?
- You are graphing a quadratic function. Does the graph change more if you change the coefficient of x^2, the coefficient of x, or the constant by 1?

Questions you might ask. Consider asking teachers questions like these:

- How are you challenging your strongest students mathematically?
- Do you let your strong students work together sometimes? Do they like that?

Feedback to offer or steps you can take. As a leader, you might ask teachers to choose one of their strongest students and take deliberate action to provide extensions for that student. Later, they can describe the action to colleagues in staff meetings and share what worked and what didn't.

Feedback

Teachers support students by giving thoughtful feedback during or at the completion of a task. There is much evidence that giving good feedback substantially improves student achievement (Hattie & Timperley, 2007). Thoughtful feedback always involves responding specifically to what the student has done or said and is rarely generic or evaluative; it focuses on the work and the learning goal, not on the student. Effective feedback guides students explicitly to ways to improve their work. This feedback might be a statement, such as "I think that, in the parts where your language was simpler, it was easier to follow your thinking," or it might be a question ("Why did you decide to make that assumption? Do you think it is justified?"). For example, if a student has written $\frac{1}{3} + \frac{3}{4} = \frac{4}{7}$, a teacher might say, "I notice your answer was a bit more than one half. Tell me why you think $\frac{1}{3} + \frac{3}{4}$ should be a bit more than one half, without actually finding the sum." This contrasts with pointing out the student's error of adding numerators and adding denominators. A principal should observe or converse with teachers about how they anticipate incorrect solutions and how to deal with them.

Many teachers give feedback that is more about whether what is written is easy for them to read (feedback that is about making the teacher comfortable) rather than about the student thinking. This might well lead students to focus on pleasing the teacher rather than on the thinking. As teachers get more comfortable with assessment criteria, their feedback should start addressing lagging concepts.

Questions you might ask. Consider asking teachers questions like these:

- I notice that you expressed concern about how easy it was for you to read the student's paper. I realize that it is an issue, but should that be the most important issue for her?
- How does the student's personality/style affect your feedback?
- How does the student's past performance affect your feedback?
- Do you feel that your feedback is helping the student improve? How?

Feedback to offer or steps you can take. As a leader, you might share a piece of student work and have teachers talk about the feedback that they would give. Point out the difference between feedback that recognizes and advances student thinking and feedback that is about organization. You might also talk to students about how they feel about the kind of feedback they are getting.

Highlighting the Important Math

An extremely important way to provide learning support for a student is for the teacher to consolidate a lesson, highlighting the important mathematics covered. There are a variety of ways to do this, but students benefit when the teacher helps the students to "hear and see" the important math ideas that were meant to come out in the lesson (Literacy and Numeracy Secretariat, 2011; Smith & Stein, 2011). For example, if a teacher is teaching about converting units of measure (e.g., feet to yards), the focus should be less on the calculation (which a calculator or app can do), and more on ideas such as why there are fewer units, why the number of yards is less than half the number of feet, and when you might want to perform a unit conversion. Instead of asking students to make a straight conversion, a teacher might pose a question like "How else could you describe how far away 5,000 days is?" or "Which of those descriptions do you think is the best and why?"

The final part of the lesson would be designed to move the student away from the particular problem that was solved toward the greater learning goal that the problem was designed to meet.

Questions you might ask. Consider asking teachers questions like these:

- I noticed that you talked with the students about the solution, but how do you know that the learning goal was brought out?
- How do you ensure that students walk away with the most important ideas you want to come out of the lesson?
- Have you considered asking students to highlight the math they learned?

Feedback to offer or steps you can take. As a leader, you might lead a discussion about the critical importance of pulling the important math out of student work and not running out of time before that is done. Some leaders insist that the teacher set a dedicated time period of 10–20 minutes per lesson to ensure that important math is brought out. You might also work with teachers on how to ask questions at the end of the lesson that focus on the math, rather than just the particular problem solved.

Summing It Up

This chapter detailed what the culture in a math school should look like when observing teachers during instruction. There has been attention to improved use of learning time, lesson focus, demonstrated confidence in students, and provided learning support for the full range of students. These are detailed in Appendix B. The chapter also offers suggestions for questions you might ask of teachers and feedback you might offer teachers to improve these facets of instruction. The next chapter focuses more on interactions with teachers outside the classroom.

6

Looking at Teachers for Signs of a Strong Math Culture Outside the Classroom

Much of what a principal might wish to learn about a teacher's instructional beliefs can be observed during instruction; however, conversations, whether quick ones in the hallway or scheduled meetings, provide valuable predictive insight into how effective teachers are likely to be in mathematics instruction. This chapter explores the information a principal receives when talking to teachers outside of the classroom and offers suggestions for actions the principal might take based on those conversations. What you might hope to hear when listening to the teacher outside the classroom falls into five categories:

- Knowledge of and attention to curriculum.
- Reflection on practice.
- High expectations.
- Attending to students' instructional needs.
- Professionalism.

Knowledge of and Attention to Curriculum

There are three elements of attention to curriculum that will be explored here:

- A deep understanding of the local curriculum.
- Appropriate resource choice.
- Attention to both curriculum content and practice standards in instruction and evaluation.

Deep Understanding of Curriculum

Teachers vary considerably in their knowledge of the curriculum or standards required in their jurisdiction. Some teachers may be less knowledgeable than others because they moved to a new district or grade level and simply have not had enough time to learn what they need to know; others, however, choose not to invest in becoming familiar with their curriculum and base what they teach either on their own history as a student or what someone else tells them to do. They sometimes do not realize that the resource they are relying on may not teach exactly what the curriculum indicates should be taught.

What might it sound like or look like? Teachers vary in their understanding of the curriculum. At one end of the scale are teachers who simply depend on their textbooks to figure out what to teach; they assume the text that they use ensures that the curriculum is covered, which may or may not be accurate. Other teachers are knowledgeable about some of their own grade's standards but not all of them, while others are more familiar with their own grade's standards but not the standards for preceding or succeeding grades. At the other end of the spectrum are those who have a deep understanding not only of their own standards but also of where they fit in the larger continuum.

Teachers need to understand the importance of learning about *all* the math standards. To help students who are struggling, teachers need to know what comes before their grade level; to ensure that they are taking appropriate direction with the topics in their grade level, teachers need to know how the topics will develop in succeeding levels. Therefore, it is important that all teachers be moving toward a deep understanding of the full range of standards.

Questions you might ask. To gain a sense of where your teachers are in terms of a deep understanding of curriculum, you might ask questions such as these:

- What aspects of the curriculum for your grade do you find "uncomfortable"? Why?
- How comfortable are you with the curriculum "continuum" from grade to grade?
- What parts of the curriculum do you think your students are not ready for?
- What parts of the curriculum do you think are missing to prepare your students for the next grade?
- How have you planned your pathway through the curriculum? What topic are you addressing first? Which topics are revisited? How?

Feedback to offer or steps you can take. As a leader, you might group teachers in your school for focused discussion. The topics could either involve "deconstructing" standards and really understanding what is required within a standard or developing shared continua that highlight the growth of topics over a grade span. For schools in the United States, these might be built on the Common Core trajectories (Common Core Standards Writing Team, 2013).

Resource Choice

The types of math resources available to teachers vary greatly. Some provide professional support for teachers; others are designed for use with students. In some jurisdictions, resource choices are made for teachers. Increasingly, though, with the advent of digital and open-source resources, teachers are using materials that they choose themselves. Resources for students might be in the form of traditional workbooks, print or digital, where students simply reply to knowledge questions, or they might be resources that truly promote mathematical understanding, as discussed in Chapter 4. It is vital that the resources teachers choose for mathematics support student understanding, rather than rote learning.

What might it sound like or look like? One problem with teachers putting resources together themselves is that it often leads to less coherent programs than those that were previously created through complex, vetted processes. Also, this practice often leads to an approach that is not sequenced appropriately. Administrators need to observe the resources that are being used and how they fit the curriculum.

One principal was in a district where textbooks were not mandated. Her approach, which was very successful, was to look more at how staff managed the curriculum. When she observed, often in class visits, that many of the lessons contained material from popular websites and social media, she worried a bit. Although engaging for the students, the principal noticed the lack of coherence and progression across grades and within grades. When monitoring, she would ask students questions about concept sequence and how the knowledge related from one day to the next. When staff asked her to visit to see the excitement of the students, she would ask questions about the lesson goal and how it related to next steps. What she found was astonishing. The lessons were chosen by popularity on social media or by student engagement; the learning was often a "hoped-for" by-product. Intentional building of foundational ideas along a continuum or progression was not easily available to the staff, as they had to use their time to find "stuff" for the students to do.

In a move to sustain excitement by students but still establish the foundational concepts in a sequence of study, the principal worked with the staff on pathway planning. With the help of knowledgeable others and curriculum personnel, she sought to move from lesson *seeking* to lesson *planning*. She began the work in grade-level groups with guiding questions:

- What's your path through the curriculum?
- How does your path help you build upon foundational concepts and models?
- How are these key concepts addressed early in the pathway?
- How does it connect to key ideas?
- How does the pathway help you differentiate?
- How do you know your sequence is working?

The result over time suggested that lesson choice is more intentional when teachers realize that sequence matters.

Questions you might ask. When teachers request support for resources, you might ask questions such as these:

- How consistent is this resource with our standards?
- How does the resource support a focus on understanding math?
- How does the resource support problem solving?
- How consistent would teaching from this resource be with what your colleagues in the school are using?

Feedback to offer or steps you can take. As a leader, you might encourage consistency in resources used in the school to ensure that students meet as coherent a program as possible. You might also guide staff to pay more attention to lesson sequencing and intention, as described above.

A Focus on Practice Standards as Well as Content

As mentioned in Chapter 1, the Common Core State Standards in the United States and all Canadian curricula include both content standards and practice or process standards. The practice and process standards focus on behaviors students should demonstrate when engaged in mathematics. These processes require explicit attention from teachers during instruction, although many teachers who are content-focused do not think deeply about the practice or process standards.

What might it sound or look like? Teachers who focus on the standards for mathematical practice or the process standards ensure that students have many opportunities to focus on problem solving and reasoning. In particular, they seek to provide opportunities for students to construct arguments, critique the arguments of others, and model real situations with mathematics. These teachers spend significant planning time considering how to set up appropriate opportunities for mathematical argumentation and mathematical modeling, like the examples shown in Figure 6.1.

Questions you might ask. In discussing with teachers their planning process, you might ask questions such as these:

- I'd love to see the students making mathematical arguments. When might I be able to come in and watch?
- How do you make sure your students have lots of opportunities for modeling real-life problems with math?
- Which of the practice or process standards is your favorite? Why?

Feedback to offer or steps you can take. As a leader, you might encourage schoolwide mathematical debates for students at the K–2, 3–5, 6–8, or 9–12 levels. These could be recorded, shared, and discussed. For example,

Figure 6.1 | **Argumentation**

Conjecture	Examples	Counterexamples
There are fewer ways to add two numbers to get to a certain sum than to subtract to get to that same amount as a difference.	$5 = 5 + 0$ or $4 + 1$ or $3 + 2$ *but* $5 = 5 - 0$ or $6 - 1$ or $7 - 2$ or $8 - 3$ or $9 - 4$ or . . .	If all rational numbers are considered, there is an infinite number of ways to obtain any value, whether a sum or a difference.
Every time you multiply, the product is greater than the factors.	$3 \times 10 = 30$ The product is greater than either 3 or 10. Similarly, $5 \times 6 = 30$ $6 \times 4 = 24$	$0 \times 1 = 0$ $1 \times 2 = 2$ $\frac{1}{2} \times 6 = 3$
The algebraic expression $3x + 4$ always has a greater value than $2x + 3$.	When $x = 2$, $10 > 7$ When $x = 10$, $34 > 23$ When $x = 100$, $304 > 203$	When $x = -1$, $1 = 1$ When $x = -10$, $-26 < -17$

2nd graders might debate whether it is more important to know how to add or subtract and why. Fourth graders might debate whether a different symbol should be used for a division situation when you are counting how many groups of one number are in another than for a division situation involving sharing. High school students might debate whether it makes more sense to solve a system of linear equations graphically or algebraically.

You might model the importance of reasoning by going into a class and asking questions to students that elicit reasoning, such as "How can this be possible?" or "Will that strategy work for all examples?" Principal questioning during visits has a powerful impact on culture, as it can reinforce the professional learning and monitor the impact. Here is what two principals we worked with told us.

> As I began to walk through our classes, I felt much more comfortable with what I wanted to see in the classroom. We had been working as a whole school on open number lines and the progression from number path to open number line to comparisons with double number lines. I noticed that paths, number lines, and clotheslines began to appear in all classrooms.
>
> When I asked staff for examples of how number lines are being used to connect to mathematical ideas, I had a variety of responses. Teachers would show videos in the staff meeting of students using a mental number line to solve problems by partitioning a number into thirds and quarters and halves.
>
> Another unusual example came in a language arts class. The teacher recounted an activity where the character in the passage exhibited both selfless and selfish traits. She put both traits as the ends of a number line. The students referred to different passages from the text on sticky notes and decided mathematically where to place the examples on the number line. In essence, the student used mathematical reasoning to analyze literary text. The teacher said to all staff that she realized that any time there is a continuum or difference in opinions, the discussion can be connected to mathematics by using a number line model.

Reflection on Practice

Teachers vary in their willingness or inclination to take the time to reflect on their teaching decisions, especially the reasons behind their decisions. This reflection could confirm their practice or encourage consideration of the need to change their practice. The literature emphasizes the importance of a reflective teacher (Çimer, Çimer, & Vekli, 2013). It is suggested that, without this reflection, teachers become unaware of the motivations and prejudices that might negatively affect student learning. Teachers who don't reflect are more likely to blame students for an unsuccessful lesson and not reflect on their own practice, which might have actually been the problem. Teachers who don't reflect are unlikely to try new, potentially more effective approaches.

Many teachers think deeply about what they do, how effective their lessons are, which students they are reaching, and what they might do differently. No teacher has all the answers, but reflective teachers always consider and reconsider learning situations in terms of becoming even more effective. Often they will seek advice from others, whether a colleague, a principal, or an expert in the field. It is critical that you encourage teachers who do not reflect naturally to do so by engaging in conversations with them that help them develop this practice. Here are several examples of such conversations:

> Principal: So what standard are you working on? What do you specifically want students to understand beyond simply doing the calculations?
>
> Teacher: I want students to see the relationship between addition and subtraction.
>
> Principal: What process standard do you want to emphasize?
>
> Teacher: I hadn't really thought about that.
>
> Principal: Well, let's think about what you wanted to do, and we can figure that out together. Ms. T. is also working on that relationship. Have you thought about planning together?

> Principal: What stood out to you about that last math lesson you did?
>
> Teacher: The kids didn't volunteer very much.
>
> Principal: Why do you think that happened?
>
> Teacher: I don't know. Maybe it was just one of those days.
>
> Principal: Have you considered other ways of getting students involved, instead of just asking for volunteers?

Reflective teachers can explain why they do what they do. In a conversation, a teacher should be able to tell you why a certain learning goal was selected, why a certain task was selected, why differentiation was or was not used in a lesson, and why particular assignments were given (Brandt, 1993). The teacher should also be able to speak to what the students already know, what is coming next, and how this lesson transitions to what is coming.

A few examples of these sorts of conversations are shown here:

> Principal: Why did you use that particular warm-up?
>
> Teacher: I just thought the students would enjoy it.
>
> Principal: But what was your learning goal? How did you think it would help toward that?
>
> Teacher: Well, it was about fractions, and that was my lesson.
>
> Principal: What do you think students better understood after the warm-up?
>
> Teacher: I think they understood that you can multiply two fractions and get a smaller or larger answer.

> Principal: I noticed you let them work with partners. Why did you do that?
>
> Teacher: I have found that students are more likely to volunteer if they've talked to a partner first.
>
> Principal: I noticed that Ryan was reluctant when you called on him. Did you expect that?
>
> Teacher: I wasn't sure; that was why I tried. Ryan has been reluctant to participate, but has been doing better lately.
>
> Principal: I noticed in the schoolwide task that Shay and a few others had difficulty placing fractions on a number line. How did she answer your questions? What are the next steps for her and the other students?

Questions you might ask. To encourage more reflection, you might ask questions such as these:

- What notes did you make that might help you next time you teach that lesson?
- Why did you decide to have all the stations at the same level?
- How did you think that task would help students see the value of factoring?
- Are you glad that you showed them the learning goal at the start of the lesson? Why or why not?

Alternately, you might initiate a conversation about how a teacher's practice has changed over time and why:

- Is that lesson different than the one you did last year on that topic? How? Why?
- How do you feel that your practice is improving? What are you concentrating on now?
- How have you used the new learning from our last session together?

Feedback to offer or steps you can take. As a leader, you might set up situations where groups of teachers are offered communal planning time and encouraged to engage in the same or similar tasks with their students. Teachers are then asked to come back and reflect on what happened. This moves teachers along on the road to reflection.

High Expectations

Teacher expectations appear both in instruction and in assessment of learning.

High Expectations in Instruction

Teachers with high expectations in instruction teach in a way that highlights problem solving and focus lessons on meaningful learning goals connected to big ideas and learning progressions. They also actively promote the mathematical standards of practice.

What might it sound like or look like? Research has proven that when teachers have high expectations, student performance improves (Marzano, 2010). Sometimes a conversation will reveal whether a teacher holds high expectations; the teacher might talk about the quality of student thinking or express optimism in the students' abilities. When a conversation reveals a teacher's lack of confidence in the students, it is likely that students are sensing this in the classroom and reacting accordingly.

Important ideas, rather than lower-level skills, become the focus of instruction for teachers with high expectations because they believe that students can learn them. Often these goals are connected to big ideas in mathematics or learning progressions (NCTM, 2014). The effective teacher leans toward learning goals that expect students to see connections and relationships, make generalizations, and make predictions (Perkins, 1998).

A teacher with high expectations might assert that the students are able to construct viable arguments and critique the reasoning of others or that they are able to persevere in solving problems. High expectations put more of the onus on the teacher to understand the depth and scope of what is being taught so that students can go beyond rote learning.

Questions you might ask. You might initiate a conversation with questions such as these:

- What were your math learning goals this week?
- How were your learning goals connected to your last learning block?
- Do you think your students are capable of high-level goals?

- What makes you think your students are weak? How can you change that?
- Which of the mathematical standards of practice come most easily to your students? Why?

Feedback to offer or steps you can take. As a leader, conduct whole-staff discussions in which you share some of the research about the positive effect of high expectations (Dweck, 2012; Rubie-Davies, Hattie, & Hamilton, 2006). Consider creating a school mission statement related to raising expectations for all students. Model high expectations yourself in terms of your staff.

Most principals already are asking for high expectations from staff. The difficulty with the implementation, however, is often defining how those expectations are manifested in student work. Determining the agreed-upon conditions for high expectations and how to know when they are met is critical across the school. Unfortunately, definitions of high expectations differ greatly within and among schools. Consequently, the expectations in a mission statement or school goals are often wide-ranging or over-generalized. If anticipations of achievement are held against high performance criteria and effective interventions are in place for remediation, it is realistic that most students can meet higher targets. Statements such as "It's great work for them" or "He can't do any better than that" do not help one know what the student really has or has not or could or could not achieve. All staff must have a consistent vision of what high expectations are and be able to demonstrate when those expectations are realized in student work.

High Expectations in Assessment of Learning

Assessment of learning is and must be linked to instruction (Guskey, 2003), because good assessment with good feedback improves performance. There are teachers who hold higher expectations for students during instruction but are reluctant to demand much of students on any formal assessment. They believe that most students cannot be held accountable to think in this way. This is a manifestation of low expectations. Teachers who focus on lower-level performance on formal assessment will likely not raise students to the levels we wish or need, because students, who tend to be practical, see that hard work is not really what the teacher values.

Questions you might ask. You might ask teachers to answer these questions as they share assessments:

- What made you choose fewer problems and more short-answer items for this test?
- What message do you think students and parents are getting about what you think is important, based on what items you have selected for this test?
- How would you anticipate your students will answer this question? How does that differ from what you would have expected last year?

Feedback to offer or steps you can take. As a leader, conduct staff discussions and reach consensus on how much emphasis should be placed on understanding and thinking questions, rather than straightforward knowledge and application questions, or on performance assessments. Encourage at least some increase over what is currently the norm. Monitor to see that teachers are adhering to the consensus achieved.

Attending to Students' Instructional Needs

In planning for students' instructional needs, strong teachers consider the classroom conditions that different students prefer and use formative assessment to differentiate instruction meaningfully (Tomlinson, 2014). This might involve differentiating structure or differentiating instruction.

Differentiating Structure

In a strong classroom climate, teachers vary lesson format and structure, knowing that individual students respond differently. There are students who thrive in a highly structured classroom, while other students blossom in a more open-ended, divergent classroom. There are students who prefer lots of independent work and others who favor more group work. There are students who like time to figure things out before a teacher intervenes and others who feel the need for immediate support. It is incumbent on teachers to learn those student preferences and find ways to support them by using varied or flexible structures, responding to different students in different ways.

Questions you can ask. When having private conversations with teachers, you might ask these questions:

- How do you decide on the structure of your lessons?
- Do you think that students would prefer that you vary it or keep it constant?
- Do you think some students might like you to change up how you deal with groups?

Feedback to offer or steps you can take. As a leader, consider how structures should be used to teach math. In some schools, principals encourage uniformity in structures, which could involve lesson style, the use of lesson time, group work, or the speed at which teachers "rescue" students. Some teachers shy away from open-ended tasks, and some embrace them. Consider being more flexible yourself and encourage teachers to experiment with different structures, specifically attending to how individual students respond to each one. Then challenge teachers to think about how they could incorporate a variety of structures into their programs to meet different students' preferences.

Differentiating Instruction

Strong teachers use different kinds of formative assessment to influence what they teach (Hattie & Timperley, 2007). These include observation notes, formal diagnostics, and conversation notes. This use of formative assessment should naturally lead to differentiation of instruction in lessons. Differentiation could be in the tasks set for the students, the nature of the feedback offered to different students, or the level or nature of the communication that is required of the student. Differentiation should ensure that all students have access to the mathematics being taught and that students who can go further are encouraged to do so. It should be a rare situation where teachers never differentiate instruction in math; just as individual readers need to read different levels of books, math students might also need different kinds of tasks. Although some teachers feel that allowing students to use varying strategies is enough differentiation, often students also benefit from working on different tasks.

Questions you might ask. You might have observed differentiation in the classroom directly, but, if not, a conversation based on these questions might be instructive:

- How are you dealing with the spread of readiness in your room?
- What are your challenges? How might I help you?
- Does differentiation appear in the instructional piece or just in the follow-up assignments?
- Is your differentiation more about strategies students use or the tasks they are asked to complete?
- Is your differentiation more about how they show the work or the work they actually do?
- Is your differentiation ensuring that your high flyers go even higher?

Feedback to offer or steps you can take. As a leader, you might share with teachers the strategies of using open-ended questions and parallel tasks as manageable ways to differentiate mathematics instruction

(Small, 2017a). Open-ended questions, as discussed in Chapter 5, offer many ways to respond correctly. (They also work well as audit tasks in that they provide important data to school leaders.) With parallel tasks, a teacher sets two or more tasks that address the same essential understandings but differ in the complexity of the details. For example, one group might create problems with three-digit numbers, and others might create problems with two-digit numbers. You should be encouraging teachers to share ways they differentiate instruction to give colleagues new ideas.

When implementing more open-ended questions schoolwide, one principal we worked with was surprised when many high-performing students commented that they had difficulty figuring out what the teacher wanted as an answer. They were comfortable with questions that tested their process of recalling the necessary steps to provide the correct answer. The principal and teachers discovered that, as a school, they were more focused on procedural answers to problems than on providing evidence of conceptual understanding. As they worked on the open-ended questions as audits of performance, the principal collected more data on the range of responses, use of consistent models, and the conceptual thinking of the students. From the data collected, it was noted that students demonstrated their mathematical understanding, and teachers were more responsive to the student answers as evidence, rather than as something to be marked right or wrong.

Professionalism

Teaching is a profession; in fact, in some jurisdictions, the legal framework for teachers' associations is called the *Teaching Profession Act* (Alberta Teachers' Association, 2016). However, not all teachers exhibit professionalism. Professionalism means that teachers continue to learn all the time, reading or viewing professional resources, experimenting with new approaches, collaborating with others to refine practice, and taking responsibility for their results (Collins, 2001).

Professional Learning

Too many teachers, both elementary and secondary, manifest discomfort with teaching mathematics (Bellock, Gunderson, Ramirez, & Levine, 2010) but choose not to do anything about that discomfort. A strong administrator can help the staff realize that professionals find out about what they don't know and do something about it. Professionalism means the same for the principal, who must also learn new things about facilitating change in math and model new learning with teachers and students. Also, principals should know the strengths and weaknesses of staff and be looking for articles, tweets, books, and other resources to share. A learning culture is always reading and experimenting with new ideas. By making new learning available, principals model knowledge acquisition. This is not the same as handing out a book at a meeting, making everyone read it, and expecting each teacher to implement the same strategy in the same way. Differentiation works as well with staff as it does with students.

Trying New Strategies

A professional, based on reading, self-reflection, or discussions with others, is always interested in trying out new ideas or strategies (Mizell, 2010). This teacher realizes that there is always room to grow and establishes a deep-seated goal of continual improvement.

Collaborating

An important aspect of professionalism is the willingness to collaborate with others, both to learn from them and to share with them (Mizell, 2010). This is one way to ensure that good practices in a single classroom spread throughout the school. You might observe teachers collaborating when they ask for common planning times, when they invite you into a joint lesson, or when they ask for technology resources to share a lesson with another classroom.

The principal can orchestrate collaboration rather than demanding it. Timetables, school organization, common pathways, new shared purchases, joint presentations/sharing, and math teams are ways in which the principal can model collaborative methods. An indicator of an improving school would be how much talk and collaboration among staff has increased.

Knowledge breeds sharing; teachers want to share what they've learned with their students. Many times, however, success remains in the confines of a single classroom. Momentum is gained through positive results. Principals can't wait for the end-of-year large-scale data; instead, they need to seek out the immediate positive change and share the results beyond the classroom.

In one school, the principal took daily videos of students whose learning focus was on multiplicative thinking. He also asked the staff to send down samples of student work that reflected the criteria introduced in the professional development. The principal guided the students through their explanation with probing questions to highlight their reasoning and then posted a video of the explanation on the school's Twitter feed. All staff and the community saw the principal participating in the learning alongside the students. It was expected that the principal's visits would be centered around student work and the changes that occurred because of new teaching strategies. Daily evidence was gathered and shared publicly by the principal. He mentioned that he wanted the improvement message to become "so loud it bursts through the classroom door into the halls and out the front door."

Taking Responsibility

When a teacher attributes poor results in mathematics only to the quality of the students and takes no responsibility for the results, alarm bells should go off. Indeed, there are times when student history is such that the teacher has a particularly tough job bringing them up to the norms for that grade level. However, effective teachers find ways to improve student attitudes and performance, regardless of student background, because they adapt instruction to student needs.

Questions you might ask. In private conversations with teachers about their mathematics instruction, you might ask questions like these:

- What professional learning are you involved in?
- What professional websites/journals/books are you reading these days?
- Which strategies from our last session do you think will work well with your students?
- What new strategies have you tried recently? How did they go?
- How often do you work with the other grade ______ teachers?
- How often do you share what you did?
- Would you be interested in working on math strategies with a small learning group including teachers from other schools?

Summing It Up

This chapter detailed what to listen for when engaging in professional discussions with teachers. There has been attention to how reflective teachers are, whether teachers hold high expectations for students, how teachers use formative assessment to differentiate instruction, and how professional teachers act. This information is detailed in Appendix C. The chapter also offers suggestions for questions you might ask of teachers and feedback you might offer teachers to improve these facets of behavior. The final chapter of this resource focuses on monitoring, evaluating, and sustaining the changes discussed in the preceding chapters.

7

Monitoring and Sustaining Improvement

This chapter will explore practices to sustain change based on data collection. The specific focus will be on how to monitor change, document change, and evaluate change in the mathematics performance of a school.

Preparing to Monitor and Document Change

Change is contingent on monitoring. All teachers, students, and parents should know that this is the school focus. This means that the principal's announcements about math; math on the school website, on social media, and in newsletters; and invitations to parents to participate in math at the school are all valuable. Highlighting math thinking in announcements is a powerful motivation for students when their work is highlighted. A particularly valuable approach for monitoring is the interactive bulletin board.

Interactive Bulletin Boards

An interactive bulletin board allows students, teachers, and even parents to post and respond to relevant work (see Figure 7.1). Anyone can contribute to an interactive board (Duff & Tonner, in press).

Figure 7.1 | **Interactive Bulletin Boards**

NUMBER LINE LANE

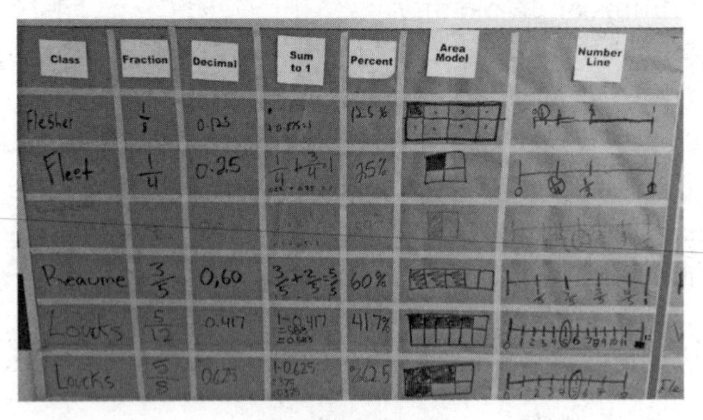

FRACTION BOARD

The items on the board relate to topics that staff are working to improve and are carefully chosen to ensure they are appropriate for everyone in the school. Open-ended questions are particularly rich, as there can be wide variety in student responses. Sometimes these problems involve computation, sometimes estimation, sometimes algebra, sometimes geometry or measurement, sometimes statistics or data. An example of an interactive board is shown in Figure 7.2.

These boards foster a math culture by encouraging conversation about math (see Figure 7.3). Staff members might note misconceptions in student

Figure 7.2 | **The Answer Is 23**

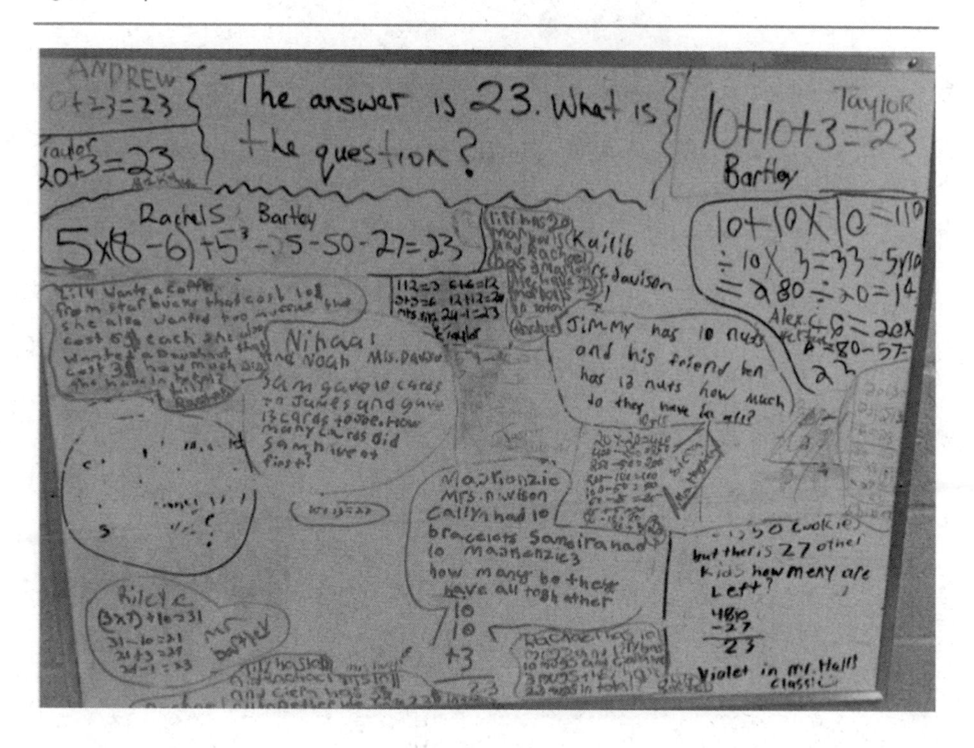

Figure 7.3 | **Math Conversations Around the Board**

STUDENTS TALK

TEACHERS TALK

work, or it might be staff or students who note growth in mathematical sophistication as students get older. The boards build community because everyone is involved. Often teachers co-learn as they stop to answer a question. Individual students could write on the board, or the board could be sectioned off for specific classes. Some classes might begin challenging other classes, leading to healthy competition. There can also be a principal's math board in a central location, highly visible, with a question or problem of the day; in some ways, the board becomes the principal's math talks, enabling the principal to follow a school plan or respond to needs seen throughout the school.

These boards build common language across the school. It is useful for students when teachers across grades share mathematical vocabulary and foster a coherent curriculum in the school.

Because they are so highly visible, interactive bulletin boards are an effective monitoring tool (Duff & Tonner, in press). The principal can note responses offered by individual students, look at a class profile, or see if a certain class is not offering any solutions and needs a bit of a push. The principal can speak to individual students or groups about ideas they have offered or go into classrooms to extend the work shown on the board or audit the quality of a class's work. The principal can also highlight key pieces from the interactive boards on the announcements or daily video.

Often the principal is the only one in the school with a panoramic view of math in the school; by using these bulletin boards, others start seeing that view. The effect goes beyond the few original champion teachers to the whole school community.

Figure 7.4 shows some teachers' observations about student reactions to these boards in one school. One student said, "They have fields for kids who like soccer, and now we have boards for kids who like learning math."

Figure 7.4 | **Reactions to Interactive Boards**

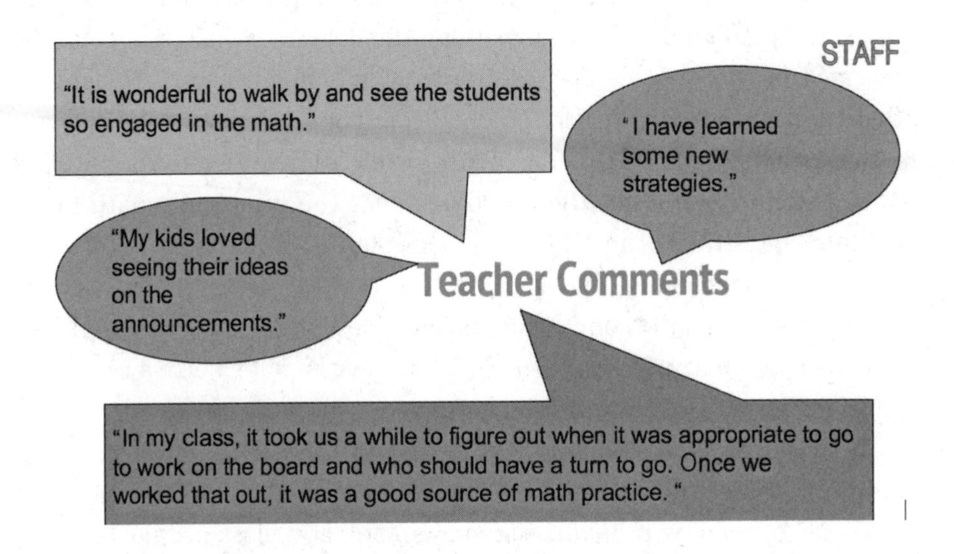

Possible Monitoring Actions

Broadcast improvement in student work to encourage others. One way to broadcast improvement is literally, on the daily announcements. A student who has exhibited good work shares it on the announcements for others to hear. Another way is through parent communication, and yet a third way is by posting work in a very visible way for others to see.

Be a part of developing the analysis criteria for common tasks. As new common tasks are developed, the principal needs to be part of the team that helps develop analysis criteria. Even though there are many other calls for a principal's time, this is not a task to delegate to others. A principal must be visibly involved.

Look for anchors on classroom walls incorporating new mathematical ideas the teacher has been exploring with the students. Ideally, the material seen on anchor charts should sometimes come from students and other times from teachers. Either way, there should always be lots of student work.

Ask teachers about extensions for top students and differentiation for struggling students. These conversations are usually casual, in the hallway or staff room or in some similar setting, but the teacher needs to know that the principal is still paying attention to the entire spectrum of students in the class.

Talk to teachers about specific students to ensure teachers are considering *each* student. As a principal we once worked with was monitoring improvement, she remarked to her colleagues that the staff were more comfortable with her "coming into the classroom and asking questions than they used to be." When prompted, the principal commented on the change of her conversations, both in the hallway or during a class visit.

She recalled a teacher's story at a staff meeting. The teacher said that she was the first principal who asked follow-up questions about a student's lagging criteria on the whole-school common task. The teacher had made a comment to the principal that, when her students were completing some of the problems on the task, they "really understood subtraction; they're just making sloppy errors." In a hallway conversation with the teacher, the principal inquired about the comment and stated that the next time she was in the class she would ask Jordan and Trisha some follow-up questions. When the principal came to the room, she asked the two students how they would solve 102 – 98. Both students immediately went to a standard algorithm and, with some difficulty, arrived at an answer that was greater than 102. When asked, "How is figuring out 102 – 98 similar to figuring out 12 – 8?", both students said it wasn't, because 12 – 8 uses "small" numbers.

After a conversation with the principal, the teacher asked the same question of several other students in the class. Later, the teacher stopped by the office to tell the principal that she had made assumptions about procedural knowledge and didn't consider the meanings of subtraction and the relationships between numbers. At the staff meeting, the teacher stood up and relayed the story to all the faculty, ending with the fact that she now looks to the student work for the understanding. She examines the student work to see if her students are making connections and reflects if she is focusing more on consolidating thinking than on just practicing a procedure. She concluded by saying, "When we focus on mathematical thinking and understanding, our perspective changes. I discovered that, when my students do not have a deeper understanding of operational meanings and number relationships, they may be more likely to make calculation errors or forget the sequence of steps in an algorithm."

Ask teachers questions about changes in student engagement. You might ask teachers about both positive and negative changes in student engagement. (Hopefully, there will be more positive ones.) Engagement means more than just participation; it indicates a real involvement in what is being learned.

Use opportunities when stakeholders are present to show student work. As the principal builds the momentum for change, it is important to invite other district leaders to the school to observe and walk through classes with the principal. The principal can provide colleagues with monitoring focus indicators or have visitors share immediate impressions of what they think is the focus in the school. If the principal has a "knowledgeable other" walk through classes, that individual can provide deeper insight into areas where the principal's knowledge requires more background.

At one school we worked with, many staff felt that their change efforts were being noticed by educators beyond the school, which strengthened their confidence in the process. When the principal announced that a district official was going to visit to observe the outstanding changes that are happening in the school, some teachers asked specifically that their classes be included because of what the students were demonstrating in their daily work.

Follow professional development opportunities or audit tasks with questions for teachers. You can ask how they can implement two or three key items based on that new learning.

Conduct monitoring surveys of staff, asking about improvement in one or two specific areas that have been under discussion. For instance, following a professional development session, one principal provided a monitoring form to all staff as part of the in-class follow-up (see Figure 7.5) (Duff & Tonner, in press). The teachers used the tool to highlight areas that required more learning or as a checklist to sustain the focus from the professional learning session. The principal let all staff know that the criteria would become the monitoring focus over the next learning period and that the teachers could use the indicators as a class or student tracker.

Ask teachers questions about the depth of student math talk and possible examples of that talk. One principal posted a video on a daily tweet to model the importance of student talk. After a professional learning session, to illustrate an aspect of the school goal, the principal sought out student work that represented the school's progression toward that goal. He modeled a mathematical conversation, highlighting the student talk by asking the student a series of questions. He felt that it was important for the teachers to see the results of their teaching reflected in the student voice.

- Tell me about the task.
- Explain why you thought . . .
- How did your work connect to what you were taught?
- Describe some of the mathematical concepts.

During the short video, he sometimes included some prompts such as "How do you know that?", "What would you do if?", and "Could you use another model"? The intent, though, was to highlight student voice as a method to support common math language and concepts and to present the idea that "math is cool at this school."

Ask teachers questions about how new ideas will be used in the next teaching cycle. Have them come up with concrete and actionable items.

Figure 7.5 | **Indicators for Monitoring**

Content/Pedagogy/Processes		
Students	**See**	**Comments**
Represent the magnitude of numbers within benchmarks		
Establish key benchmarks beyond 5 and 10		
Extend from benchmark applications		
Estimate using multiple benchmarks		
Deeply decompose numbers beyond place value		
Demonstrate understanding of embedded quantity		
Apply missing-number situations in other strands		
Check for reasonableness with another model		
Use the math tool to show understanding		
Make number connections across strands		
Double, halve, triple, and third numbers		
Classroom	**See**	**Comments**
Intentional focus on numbers chosen		
Explicit teaching of math concepts through concrete/representational/symbolic forms		
Set models, linear models, area models represented in student work and anchor charts		
Explicit evidence of properties within addition and subtraction: commutative and associative		
School	**See**	**Comments**
Staff believe they make the difference in learning mathematics for students		
Student work guides professional learning needs		
Ongoing monitoring of teaching/learning occurs, using criteria		

NOTES:

Source: From *MathUP School: School Improvement Cycle for Mathematics*, by D. Duff and J. Tonner, in press, Oakville, Ontario, Canada: Rubicon Publishing. Copyright 2018 by Rubicon. Reprinted with permission.

Sustaining Change

Evaluating change, of course, depends on collecting data. Collected data focus on the students, the teachers, and the principal. The principal's mindset required for sustaining change is one of relentless belief that a new potential is attainable. This attitude is not a defined initiative but a call to action that requires everyone to have a deep understanding of the role of data in instruction and implementation.

- *Focused intentionality* for school improvement means that the principal leads the data collection across the school, sets the standards for improvement, and connects or discards competing initiatives.
- *Urgent patience* allows all staff to feel they have time to use their data to go deeper, but at the same time there is an urgency for change. Students must be affected by our new knowledge *today*.
- *Regulated accountability* asks everyone to be held to a standard for growth. It is not a tiered approach. Everyone must feel invested in the success, and, individually, everyone's part is examined regularly (Duff & Tonner, in press).

Data Focused on Students

To show real improvement, student work must better meet analysis criteria associated with audit tasks than criteria met in earlier work. Leaders should also see more look-fors from Chapter 4 that indicate increased student engagement and deeper understanding of math. This would normally involve better and more varied representations, better and more varied use of strategies, and better communication of mathematical reasoning.

Data Focused on Teachers

The look-fors in Chapters 5 and 6 are great indicators of improved practice in math instruction. It is particularly critical to gather data in these areas.

Whether teachers make good predictions about how both individual students and whole classes will perform (Duff & Tonner, in press). These predictions would likely have been made prior to examining schoolwide and audit tasks or in hall conversations with the administrator.

Just as it is critical to have students predict outcomes, it is important for teachers to predict what the student growth should be after a teaching cycle. One of the markers for sustaining high-quality student work is for the teachers to understand the effect they have on the performance of a student in mathematics.

In the first month at a new school, a principal we worked with conducted an informal survey of teachers on how much impact they thought their teaching had on their students' scores. The principal concluded that, in general, the staff felt that they had more impact on student's reading and writing improvement than on mathematics. When addressing this dichotomy with the staff, the argument from the staff was based on past evidence. The staff continually witnessed their impact on student ability to read and comprehend factored into district scores, yet, in math, they felt that the students "either get it or they don't." In describing change in literacy, teachers pointed to the teaching, but changes in math growth were attributed only to the student's innate abilities ("They will be good despite my teaching").

In the school, there were identifiable markers for student growth in reading and writing, but when teachers were asked what a year's growth looked like in mathematics, they discussed what needed to be covered in the curriculum instead. To change this mindset in mathematics, the principal had the staff predict student change over the course of a learning cycle.

When the staff became invested in documenting student change against key criteria, they began to see the effect of their work and knowledge reflected in the student outcomes.

Now, in the third year of massive change in district scores in mathematics, the principal sustains the progress with continual focus on the assessment/instruction connection. The teachers continually discuss student work compared to essential understandings, and they constantly refine their predictions for each student over the course of several teaching blocks.

Whether teacher's indications of what they think a year's growth should look like actually show up in student work. These indications would likely have been made after examining schoolwide and audit tasks.

Figure 7.6 (Duff & Tonner, in press) shows how one school determined what a year's growth looks like and how to sustain the changes that had been made in individual grades. After a working session with kindergarten teachers, the principal asked, "Now that we made some significant changes in student performance, how do we sustain that change from year to year and teacher to teacher?" Because of the focused work the kindergartners had been doing, the students were demonstrating many important concepts in their early number understandings. When the principal asked the teachers to predict how many of their students could perform at the highest level on a district test, based on these early number principles, each teacher easily identified two to four students. The total prediction was slightly above the district average. The principal noted that, on the 3rd grade large-scale district test, no students placed in the top category. What happened to those students in mathematics over the course of three years? The Primary Number Markers was a tool this school developed to track and sustain student growth over several years. It could be used to track individual students, and it also served as a class snapshot over the course of a year.

Figure 7.6 | **Primary Number Markers**

Student:			**Date:**		
Number Relationships	**Description**	**Notes**	**Has It**	**Almost**	**Not Yet**
Comparing: more, less, equal					
One more/one less					
Two more/two less					
Difference, comparison					
Odd/even beyond recognition					
Properties: commutative ($a + b = b + a$), associative ($[a + b] + c =$ ß $a + [b + c]$)					
Pairs that make 10 (3 + 7, 4 + 6, 5 + 5, . . .); also missing number (4 + ______ = 10)					
Recognizing missing number situations ($a + \square = c$)					
Division (fair share)					
Fractional connections by sharing					
Two times/doubling, halving, tripling					
Number Lines/Paths	**Description**	**Notes**	**Has It**	**Almost**	**Not Yet**
Uses number lines with benchmarks					
Skip counting by 2, 5, 10, 3 (not always starting at 0)					
Addition					
Two meanings of subtraction (take away, adding up)					
Proportional spacing/jumps (five times bigger than 1)					

Figure 7.6 | **Primary Number Markers** *(continued)*

Decomposition	Description	Notes	Has It	Almost	Not Yet
Doubles, halves					
Base ten					
Benchmarks (on/off)					
Decomposing additively					
Unitizing (multiplicative)					
Part/whole (*ABC* model based on $a + b = c$ or $a - b = c$)					
Adding and subtracting by decomposing					
Benchmarks	**Description**	**Notes**	**Has It**	**Almost**	**Not Yet**
2, 5, 10, 12					
15, 20					
24, 25					
Halves, then fourths					
Thirds, eighths					

Can Use Manipulative/Tool to Represent

Manipulative/Tool	Y	N	Manipulative/Tool	Y	N	Manipulative/Tool	Y	N
Linking cubes			Coins			Dominoes		
Base ten blocks			Tiles			Measuring tapes		
Ten frame/five frame			Cuisenaire rods			Number cubes		
Hundreds chart			Double-sided counters			Pentominoes		
Rekenrek			Pattern blocks					

Source: From *MathUP School: School Improvement Cycle for Mathematics*, by D. Duff and J. Tonner, in press, Oakville, Ontario, Canada: Rubicon Publishing.

Whether audit tasks and criteria make sense. Often teachers' criteria focus primarily on correctness, organization, or numbers of strategies used, rather than on the sophistication of models, the appropriateness of the strategies (not just how many), and deeper understanding of the foundational ideas. For a school to sustain the momentum of positive change, audit tasks and criteria should become part of a regular routine for students. Furthermore, criteria must look not only at procedures but also at deep mathematical concepts and misconceptions. In our work at several schools, teachers found that it was difficult to analyze overly general criteria with any consistency. Student self-assessment statements often looked like this:

- I try my best and persevere.
- My solution is organized.
- I check my solution for reasonableness.
- I use the most efficient strategy.

In many schools, criteria actually appeared to be targeted toward literacy, rather than mathematics:

- I read the problem carefully, and I underline the key words.
- I write a complete solution in sentences.
- I include a math sentence.
- I explain my thinking using math words.
- I explain how my solution is correct.
- My math diagrams are labeled.

Instead, it is essential to develop criteria that are mathematically specific and sustain the action research intent (Duff & Tonner, in press). They can be used to see the following: How does the teacher change classroom teaching as a result of the criteria? How does the students' work change according to the criteria?

The indicators in Figure 7.5 show how the principal can model the intent and specificity of the criteria. If the principal's "class" comprises the teachers in the building, the "class" can see that task and criteria design must be relevant and target big ideas.

Data Focused on Principals

The administrator must honestly assess his or her own work in these areas.

Whether he or she has built collective consensus in a positive direction for math change. Although consensus does not mean uniformity, it should mean that the school community is consistent in emphasizing understanding over procedures, encouraging growth in strategies and models used, and empowering students to believe they can be successful.

Change in just a few classrooms is the enemy of sustained school improvement. Many school leaders inadvertently judge their school's growth on the data from a few classes and generalize those data to all the classrooms in the school. These frontrunners and early adopters bring the first wave of the improvement. Upon realizing improved indicators, these teachers are the first to show their data, discuss student work, and share during professional learning. When analyzing change, these champions come immediately to mind and possibly bias the data across the school. Staff find themselves using generalized statements such as "We are really making a difference in our math" or "Implementing our new initiative has changed the way we look at teaching math." If schools are not reaching second- and third-tier adopters, they will only be able to sustain that pocketed change (Duff & Tonner, in press). Coherence and collective effectiveness involves all staff members. When principals are leading intentionally, actively participating and learning alongside staff, their assessment of change includes all staff, as well as their own learning.

Whether he or she has collected data to ensure appropriate follow-up for individual teachers. These data will involve more than just audit tasks; they will include classroom work that gets better and better.

Whether he or she has been so clear about the vision that even students can articulate it. With many school improvement initiatives, building a vision is often a first step in the process. Many schools display their vision proudly in the lobby. However, once the vision has been established, there is often another initiative that comes along to take its place, and the momentum is lost. Schools that are sustaining positive growth live their vision daily, and both staff and students can articulate the main goal of the school. The proof of sustaining the belief is seen in the students, not in how the goals are posted or where they are written.

There are different ways to assess the clarity of the vision. When the principal or visitor asks a cross-section of the students what the school is about, do the students respond with a version of the vision or goal of the school? Do the students reflect the efforts of the staff toward coherence? How do students generally perceive the school? When the visitor asks the staff, do they respond similarly?

One principal we worked with charted the narrative of her school as it was moving through its change cycle. Despite starting with the vision and goals around mathematics, the principal recalled that, in the first six months, the responses from the students did not change significantly. Comments like "We are a ghetto school," "We are one of the worst schools around," and "No one really cares about how we do" were common. No amount of paint, cleaning, or posted mission statements changed the students' perceptions. The principal stated that, when assessment scores increased as teachers' practice deepened, the focus of the student narrative turned from the environment and was directed more toward teaching and learning. Each time a district visitor would come to the school, that person would walk with the principal through the halls and classrooms and ask the students, "What is this school about?" One of the top three responses was always something close to "We are a math school!"

Whether he or she has led staff in intentional use of data to analyze change. One principal we know tells this story:

> "For our teaching staff, it was the analysis of the data that changed preconceived notions on the impact of their teaching. It also completely revamped my working with data analysis. We knew from our scores that our students were not meeting the district standards. We didn't realize the extent of the issues until we analyzed the student data.
>
> "Working with my principal coach, we were discussing the importance of fractional magnitude on later mathematics success (Siegler & Pyke, 2013). When I discussed the research and the concerns with some of our staff, it was clear that they felt their students understood fractions because of their teaching and their 'strong fractions unit.' After some debate, we all agreed that we would attempt a focused task on the use of the number line and fractional magnitude. In their prediction, the teachers thought that most of their students would meet the criteria outlined on our Criteria for Analysis Tracker. The staff believed the students would be able to order fractions from least to greatest, would understand the meaning of the numerator and denominator, and would understand fractions greater than one. What appeared in the data shocked us all and transformed the intentionality in the way we teach and how we use data analysis. Most of our students showed huge gaps in the criteria, and the results suggested little understanding of fractional magnitude. All students except two did not place $\frac{4}{3}$ accurately on a number line, and more than 80 percent stated they ordered the fractions by the 'first number' (size of the numerator) (see Figure 7.7) (Duff & Tonner, in press).

"When we analyzed the misconceptions, our preconceived notions immediately went from 'what I taught' to 'what I see in the work.' The teacher team presented their findings at our next faculty meeting and told the staff how this work has changed their thinking about student data."

Figure 7.7 | **Fraction Math Task**

Order the following 5 numbers from least to greatest. Use a model to support your answer.

2/5, 4/3, 5/6, 1, 1/7

1, 1/7, 2/5, 4/3, 5/6

I think that would be least to greatest because the first digits of the fraction are bigger then just 1.

Source: From *MathUP School: School Improvement Cycle for Mathematics*, by D. Duff and J. Tonner, in press, Oakville, Ontario, Canada: Rubicon Publishing. Copyright 2018 by Rubicon. Reprinted with permission.

Whether teachers show growth in making more informed math teaching decisions. A principal visiting a class should begin to see evidence of planning decisions that include newly learned concepts. Soon the decisions move beyond the classroom and involve more teachers in the school. The school growth can become exponential as the changes move beyond one teacher's class.

In one case, a principal we know observed that staff planning had changed. Before the change, the teachers' planning was a solitary process guarded by individuals and was a badge for those who prided themselves in their singular effort. The only person who saw the plans of all staff was the principal.

As they began to work on common models across the school with success, the teachers realized early in the following school year that their students had already internalized and applied the mathematical models taught in the previous grade. Planning pathways through the curriculum became a shared effort as the teachers realized that each grade was accountable for the year's targets. The principal noticed there was much less reteaching or review of the previous grade's concepts. Staff would comment that the students' new knowledge had motivated them to collaborate and share planning with other staff.

Whether the principal, with teachers, consistently uses action research in the school to improve instruction. Establishing a research mindset within the school becomes a refined methodology to sustain positive math change. The principal's aim is to construct a school with a culture of inquiry, where reflective practice on the part of the teachers is the norm in all classes. Principals see data every day from a panoramic perspective; teachers refine that perspective daily based on their students' responses and actions. There is also an energy associated with this daily reflection and a sense of excitement that, as the principal influences school change, the teacher effects class change and the students see the effect on their own learning. How do we leverage that experience from each class to the whole school?

Six Essential Aspects of Action Research

This section summarizes essential components of the type of action research described previously in Chapters 2 and 6 and in this chapter as well.

Actions must be derived from focused professional development. The principal and the school team establish, from the start, where the needs begin based on student work.

Evidence criteria must be formulated. The principal, along with the teachers, determines what good work looks like in making evidence decisions. Exemplars are chosen to show everyone what good work looks like.

Cycle audit points must be established. The principal and the staff define when the next "snapshot" or audit point will occur. Audit points can be used to redefine the cycle timelines, refocus the instruction, or move to the next stage of the action research.

Data must be analyzed. The principal and teachers must constantly analyze the student work and instruction to support staff. They should avoid drawing conclusions from short timelines or small data sets.

The next cycle of professional development must be targeted. From the evidence analysis, target new learning around common misconceptions, concept progressions, remediation, differentiation, or other areas that need to be addressed.

Potential problems must be identified and avoided. Common pitfalls of school improvement efforts include the following:

- Moving off the focus too quickly
- Allowing too many competing actions
- Adding too many components not connected to the agreed-upon target
- Insufficient mathematics background on the part of the teachers without an attempt to learn what is missing
- Inexperience with reflective practice

A major benefit of action research is that it allows a principal to move away from traditional programs that have no evidence of change over time. Through excitement at seeing the change in student achievement, teachers see a benefit in their new work and are more likely to monitor effectiveness continually.

The administrator can also look at his or her own work on aligning school life to the vision in terms of how resources are allocated; whether the timetable reflects and supports the vision; and whether announcements, bulletin boards, and assemblies are related to the vision. Positive growth is difficult to sustain without a critical mass. The voice of change begins with the principal and is picked up by the staff, who amplify the message. The principal develops the depth and spread that is needed through an unwavering commitment to monitoring progress in mathematics.

Complacency is the enemy. If the principal deviates from the mission by not connecting the communication, timetabling, and resources, the inherent message to staff is that "you say *this* but it's really about *that*." Principals should ask themselves some tough questions:

- What's my *that*?
- How is the focus connected to every large move I make?
- How am I keeping the main message relevant to all stakeholders?

All staff understand that in any school there are many competing initiatives, important programs, and daily interruptions. They know all the work is important; they won't abandon the teaching of reading, for example, just because the focus is math improvement.

The importance of sustaining the mathematics mission became overwhelmingly obvious to a principal we once worked with. She had moved into a new school that was showing tremendous improvement in student mathematics scores and wanted to include a few other initiatives alongside the mathematics work. The teachers told the principal that their staff's biggest fear was losing the momentum by changing or diluting their focus. For the first time, the school was building momentum in math, and it had given the staff and students a confidence that they didn't want to lose. The change had even affected student behavior and literacy scores.

The principal stated that the staff was far ahead of her in terms of mathematics knowledge, and she was worried that she had little to offer, based on her background. However, never had the experienced principal felt that her effectiveness or ability to impact a school was being challenged. As she tells the story, the best decision she ever made was to relinquish her need for control and begin learning alongside the staff, using her leadership skills as added value. She forced herself to examine the practice in the classes, immerse herself in the students' work, and learn together with and from the staff. The realization that she recounted at numerous leadership meetings was simple: "I knew the work in mathematics was transformational, so, if they can do it, I can lead it—but, first, we will work it together."

Summing It Up

This chapter explained how improvement for a school might be sustained, monitored, and evaluated. An important tool is the interactive bulletin board, which can be used for monitoring and documenting change while maintaining attention to the school's focus on improvement in mathematics. There were also detailed suggestions regarding data about students, teachers, and the principal that should be collected to monitor and evaluate the success of the change.

An Overall Summary

If there is one message that we would like to leave you with, it would be that individual teachers might do amazing things, but without school leadership focused on building math success across the school and changing the culture of the school, there will not be school or system change. School leaders, whether they are personally comfortable in math or not, must be involved if their schools are to change their students' success with math and sustain that change.

We have tried to provide you with detailed information about how to facilitate change, what to look for to see how it is working, and how to sustain that change. We know it works; we've done it.

APPENDIX A

Look-Fors, Questions, and Feedback for Students

	What You Are Looking For: Students
Rich math talk	Students are provided with opportunities to engage in rich math talk (e.g., debates). There is more student talk than teacher talk. The math talk is real math conversation.
Collaboration	Students regularly collaborate. Students seem comfortable when they are collaborating. Students have been taught what collaboration looks like.
Questions asked by students	The questions students ask are mostly to clarify, justify, and challenge thinking (not about organization).
Student use of learning tools	Students are using manipulatives or technology in a rich way, not just procedurally.

	What You Are Looking For: Students
Student engagement with math	Students seem enthusiastic about what they are doing because the tasks are engaging. Students regularly seek to make sense of what is going on and do not ask for quick, easy answers just to get finished. Students comfortably explain their thinking.
Student confidence	Students feel confident trying things and do not feel the need to seek teacher direction or approval. Students show perseverance in situations where perseverance is appropriate. Students seem comfortable and proud of being creative and do not feel the need to please or conform in terms of their math thinking.

	What You Are Looking For: Principals and Teachers
Rich math talk	Teachers are provided with opportunities to engage in rich math talk (e.g., hallway conversations, interactive mathboard, student work). Staff meetings are times where teachers model rich math talk around problems, student work, pedagogy, and new ideas.
Collaboration	The principal promotes and models regular math collaboration. The principal creates situations where teachers collaborate with other teachers who teach a different grade. Math conversations occur at many times in the school, both informally and formally.
Questions asked by teachers	The questions teachers ask the principal are beyond administrative and clarify, justify, and challenge thinking. Questions often revolve around student work.
Learning tools	Teachers' use of manipulatives or technology is increasing across the school. The principal asks students questions about how they use manipulatives and technology. The principal models manipulative work at staff meetings.
How principals engage with the math	The principal models enthusiasm about math improvement. The principal consistently speaks to students about their math learning. The principal supports student sense-making and regularly asks teachers how reasoning and proving is visible in the instruction.
How confident principals are with math instruction	Teachers are comfortable discussing math and math instruction with the principal, and the principal seeks out math conversations. The principals is often observed co-learning with the teachers. The principal supports teachers' creative endeavors in math. The principal supports action research when trying creative processes.

APPENDIX B

Look-Fors, Questions, and Feedback for Teachers in the Classroom

	What You Are Looking For: Teachers
Good use of learning time	The teacher avoids wasting unreasonable amounts of time on activities like the following that do not focus on the learning goal: • Management issues • Homework review • Activities not appropriate to particular learners • Too much communal sharing
Lesson focus	The teacher frequently teaches through problem solving. The teacher poses problems that pique curiosity or lead to generalizations. The teacher focuses questions and tasks on understanding and thinking, rather than just repetition of procedures.
Confidence in students	The teacher usually *asks* rather than *tells*. The teacher promotes divergent thinking.
Learning support for students	The teacher takes the time to listen to students. The teacher supports struggling students by probing and scaffolding but not telling. The student provides extensions for strong students. The teacher gives feedback based on clear criteria. The teacher consolidates student thinking in each lesson to highlight the most important mathematical issues.

	What You Are Looking For: Principals
Good use of learning time	Teachers have opportunities at staff meetings to discuss strategies they have developed to use time more effectively in the classroom.
Lesson focus	The principal models focus in developing a schoolwide math approach.
Confidence	The principal shows confidence in teachers by encouraging them to personalize the communal path. The principal shows confidence in teachers by letting them take the lead in some professional development opportunities. The principal shows confidence in students by encouraging participation for all grade levels in communal tasks.
Learning supports	The principal takes the time to listen as teachers share their teaching successes and struggles. The principal provides support by scaffolding carefully and providing useful formative feedback. Extensions are offered to even the strongest teachers.

APPENDIX C

Look-Fors, Questions, and Feedback for Teachers Outside the Classroom

	What You Are Looking For: Teachers
Reflection on practice	The teacher shows in conversation that he or she regularly considers instructional decisions and reconsiders them based on events or advice or new information. The teacher can explain his or her instructional choices; they are not "random" or simply "that's what I always do."
High expectations	There is evidence in conversations that the teacher focuses on meaningful learning goals connected to big ideas and learning progressions. The teacher demonstrates familiarity with mathematical standards of practice and can talk about how to promote them with students. The way the teacher talks about assessment reflects a focus on thinking, problem solving, and understanding.
Attends to students' instructional needs	The teacher can explain how she or he varies the format and structure level of lessons to accommodate a broader range of students. The teacher can explain how she or he uses formative assessment to meaningfully differentiate instruction to accommodate a broader range of students.
Professionalism	The teacher can speak to professional reading and learning he or she has engaged in lately. The teacher is open to and interested in new ideas. The teacher collaborates with others. The teacher shows in conversation how he or she takes responsibility for his or her students' success rather than "blames" the student.

	What You Are Looking For: Principals
Reflection on practice	The principal promotes and models regular reflection on his or her own practice or decisions. For example, he or she reflects on student data with the staff and collaborates on change practices. The principal regularly explains his or her own choices for directions chosen for the school.
High expectations	The principal exhibits high expectations in interactions with students, parents, and teachers. For example, the principal frequently asks questions to students as a result of class visits or data results. The principal collects data about his or her own effectiveness in moving the math agenda.
Attends to teachers' instructional needs	The principal varies the format and structure of interactions with staff based on staff preferences. The principal meaningfully differentiates the types of support and interaction given to various teachers based on their needs.
Professionalism	The principal is seen to be and is a learner–reading professionally and collaborating based on what she or he learns. The principal shares his or her learning with staff–connected to where teachers are on their learning curve. The principal shows interest in new ideas that are suggested. The principal collaborates with both staff and other principals. Just as teachers know at what level their students are achieving at all times, the principal should know where each teacher is in his or her learning at all times. The principal takes responsibility for the school's success or lack of success.

Further Reading

Bakker, A., Smit, J., & Wegerif, R. (2015). Scaffolding and dialogic teaching in mathematics education: Introduction and review. *ZDM Mathematics Education, 47*(7), 1047–1065. doi: 10.1007/s11858-015-0738-8

Hemphill, C., Mader, N., & Cory, B. (2015). *What's wrong with math and science in NYC high schools (and what to do about it).* Retrieved from https://static1.squarespace.com/static/53ee4f0be4b015b9c3690d84/t/55c413afe4b0a3278e55d9a7/1438913455694/Problems+with+Math+%26+Science+05.pdf

Ontario Ministry of Education. (2011). *Paying attention to mathematics education.* Retrieved from http://www.edu.gov.on.ca/eng/teachers/studentsuccess/FoundationPrincipals.pdf

Ontario Ministry of Education. (2016). *Press release: Ontario dedicating $60 million for renewed math strategy.* Retrieved from https://news.ontario.ca/edu/en/2016/04/ontario-dedicating-60-million-for-renewed-math-strategy.html

Wagganer, E. L. (2015). Creating math talk communities. *Teaching Children Mathematics, 22*(4), 249–254.

Wiggins, G. (2012). Seven keys to effective feedback. *Educational Leadership, 70*(1), 10–16.

References

Alberta Teachers' Association. (2012). *Nature of teaching and teaching as a profession.* Retrieved from https://www.teachers.ab.ca/About%20the%20ATA/What-We-Think/Position%20Papers/Pages/Nature%20of%20Teaching%20and%20Teaching%20as%20a%20Profession.aspx

Bellock, S. L., Gunderson, E. A., Ramirez, G., & Levine, S. C. (2010). Female teachers' math anxiety affects girls' math achievement. *PNAS (Proceedings of the National Academy of Sciences of the United States of America), 107*(5), 1860–1863.

Boaler, J., & Staples, M. (2008). Creating mathematical futures through an equitable teaching approach: The case of Railside School. *Teachers College Record, 110*(3), 608–645.

Brandt, R. (1993). On teaching for understanding: A conversation with Howard Gardner. *Educational Leadership, 50*(7), 4–7.

Chapman, O. (2012). Prospective elementary teachers' ways of making sense of problem posing. *PNA, 6*(4), 135–146. Retrieved from https://documat.unirioja.es/descarga/articulo/4266850.pdf

Çimer, A., Çimer, S., & Vekli, G. (2013). How does reflection help teachers to become effective teachers? *International Journal of Educational Research, 1*(4), 133–149.

Clarkson, P., Bishop, A., & Seah, W. T. (2010). Mathematics education and student values: The cultivation of mathematical wellbeing. In T. Lovat, R. Toomey, & N. Clement (Eds.), *International research handbook on values education and student well-being* (pp. 111–135). New York: Springer.

Collins, J. (2001). *Good to great: Why some companies make the leap . . . and others don't.* New York: HarperCollins.

Common Core Standards Writing Team. (2013). Progressions for the Common Core State Standards in mathematics. University of Arizona Institute for Mathematics and Education. Retrieved from http://math.arizona.edu/~ime/progressions/

Council of Chief State School Officers & National Governors Association. (2010). *Standards for mathematical practice.* Retrieved from http://www.corestandards.org/Math/Practice/

Dale, R., & Scherrer, J. (2015). Goldilocks discourse: Math scaffolding that's just right. *Phi Delta Kappan, 97*(2), 58–61.

Dickson, B. (2011, September 7). What if Angry Birds didn't grade with SBG? [blog post]. Retrieved from *Bowman in Arabia* at https://bowmandickson.com/2011/09/07/what-if-angry-birds-didnt-grade-with-sbg/

Duckworth, A. L., Peterson, C., Matthews, M. D., & Kelly, D. R. (2007). Grit: Perseverance and passion for long-term goals. *Journal of Personality and Social Psychology, 92*(6), 1087–1101.

Duff, D., & Tonner, J. (in press). *MathUP school: School improvement cycle for mathematics.* Oakville, Ontario, Canada: Rubicon Publishing.

Dweck, C. (2012). *Mindset: Changing the way you think to fulfill your potential.* London: Little, Brown Book Group.

Fletcher, G. (2016). *Making sense series: The progression of addition and subtraction.* Retrieved from https://gfletchy.com/2016/03/04/the-progression-of-addition-and-subtraction/

Fullan, M. (2006). *Change theory: A force for school improvement.* Jolimont, Australia: CSE Centre for Strategic Education.

Furner, J. M., & Berman, B. T. (2005). *Confidence in their ability to do mathematics: The need to eradicate math anxiety so our future students can successfully compete in a high-tech globally competitive world.* Retrieved from http://socialsciences.exeter.ac.uk/education/research/centres/stem/publications/pmej/pome18/furner_math_anxiety_2.htm

Gomes, P. (2016). *How PISA is changing to reflect 21st century workforce needs and skills.* EdSurge. Retrieved from https://www.edsurge.com/news/2016-04-26-how-pisa-is-changing-to-reflect-21st-century-workforce-needs-and-skills

Gordon, A. (2016, November 30). Ontario math scores for 10-year-olds slip in global rankings. *The Toronto Star.* Retrieved from https://www.thestar.com/news/gta/2016/11/30/ontario-math-scores-for-10-year-olds-lag-behind-27-other-countries.html

Green, E. (2014, July 23). Why do Americans stink at math? *The New York Times Magazine.* Retrieved from https://www.nytimes.com/2014/07/27/magazine/why-do-americans-stink-at-math.html

Guskey, T. R. (2003). How classroom assessment improves learning. *Educational Leadership, 60*(5), 6–11.

Hattie, J. (2012). *Visible learning for teachers: Maximizing impact on learning.* New York: Routledge.

Hattie, J., & Timperley, H. (2007). The power of feedback. *Review of Educational Research, 77*(1), 81–112. doi:10.3102/003465430298487

Hendy, H. M., Schorschinksy, N., & Wade, B. (2014). *Math confidence scale.* Retrieved from PsycTESTS. doi: 10.1037/t38161-000

Hufferd-Ackles, K., Fuson, K. C., & Sherin, M. G. (2004). Describing levels and components of a math-talk learning community. *Journal for Research in Mathematics Education, 35*(2), 81–116.

Kosko, K. W. (2016). Primary teachers' choice of probing questions: Effects of MKT and supporting student autonomy. *IEJME-Mathematics Education, 11*(4), 991–2012.

Liljedahl, P. (2014). The affordances of using visibly random groups in a mathematics classroom. In Y. Li, E. Silver, & S. Li (Eds.), *Transforming mathematics instruction: Multiple approaches and practices* (pp. 127–146). New York: Springer.

Literacy and Numeracy Secretariat. (2011). *Bansho (board writing).* Capacity building series. Retrieved from http://thelearningexchange.ca/wp-content/uploads/2017/02/CBS_bansho.pdf

Marzano, R. J. (2010). Art and science of teaching: High expectations for all. *Educational Leadership, 68*(1), 82–84.

Marzano, R. J., Marzano, J. S., & Pickering, D. J. (2003). *Classroom management that works: Research-based strategies for every teacher.* Alexandria, VA: ASCD.

Meyer, D. (2011, May 11). The three acts of a mathematical story [blog post]. Retrieved from *Dy/Dan* at http://blog.mrmeyer.com/2011/the-three-acts-of-a-mathematical-story/

Mizell, H. (2010). *Why professional development matters.* Oxford, OH: Learning Forward.

NCTM (National Council of Teachers of Mathematics). (2000). *Principles and standards for school mathematics.* Reston, VA: Author.

NCTM (National Council of Teachers of Mathematics). (2006). *Curriculum focal points for prekindergarten through grade 8 mathematics: A quest for coherence.* Reston, VA: Author.

NCTM (National Council of Teachers of Mathematics). (2009). *Focus in high school mathematics: Reasoning and sense making.* Reston, VA: Author.

NCTM (National Council of Teachers of Mathematics). (2014). *Principles to actions: Ensuring mathematical success for all.* Reston, VA: Author.

NCTM (National Council of Teachers of Mathematics). (2016). Beginning to problem solve with "I notice, I wonder". *The Math Forum.* Retrieved from https://www.nctm.org/Classroom-Resources/Problems-of-the-Week/I-Notice-I-Wonder/

Otten, S., Cirillo, M., & Herbel-Eisenmann, B. (2015). Making the most of going over homework. *Mathematics Teaching in the Middle School, 21*(2), 98–104.

Perkins, D. (1998). What is understanding? In M. S. Wiske (Ed.), *Teaching for understanding: Linking research with practice* (pp. 39–57). San Francisco: Jossey-Bass.

Piggott, J. (2008, September). Rich tasks and contexts [blog post]. Retrieved from Nrich at http://nrich.maths.org/5662

PISA envy. (2013, January 19). *The Economist.* Retrieved from http://www.economist.com/news/international/21569689-research-comparing-educational-achievement-between-countries-growing-drawing

Rubie-Davies, C., Hattie, J., & Hamilton, R. (2006). Expecting the best for students: Teacher expectations and academic outcomes. *British Journal of Educational Psychology, 76*(3), 429–444.

Siegler, R. S., & Pyke, A. A. (2013). Developmental and individual differences in understanding of fractions. *Developmental Psychology, 49*(10), 1994–2004.

Silver, E. A. (1997). Fostering creativity through instruction rich in mathematical problem solving and problem posing. *ZDM (Zentralblatt für Didaktik der Mathematik), 29*(3), 75–80.

Small, M. (2017a). *Good questions: Great ways to differentiate mathematics instruction in the standards-based classroom* (3rd ed.). New York: Teachers College Press; and Reston, VA: NCTM.

Small, M. (2017b). *Making math meaningful to Canadian students, K–8* (3rd ed.). Toronto: Nelson Education.

Small, M. (2017c). *Teaching mathematical thinking: Tasks and questions to strengthen practices and processes.* New York: Teachers College Press.

Small, M., & Lin, A. (2010). *More good questions: Great ways to differentiate secondary math instruction.* New York: Teachers College Press; and Reston, VA: NCTM.

Smith, M. S., & Stein, M. K. (2011). *5 practices for orchestrating productive mathematics discussions.* Reston, VA: NCTM.

Sowell, E. (1989). Effects of manipulative materials in mathematics instruction. *Journal for Research in Mathematics Education, 20,* 498–505.

State Government of Victoria, Australia. (2017). Scaffolding numeracy in the middle years: Authentic tasks. Retrieved from http://www.education.vic.gov.au/school/teachers/teachingresources/discipline/maths/assessment/Pages/authtasks.aspx

Stigler, J. W., & Hiebert, J. (2004). Improving mathematics teaching. *Educational Leadership, 61*(5), 12–16.

Stutz, T. (2013, February 21). Texas students outpace rest of U.S. in math, science, lag in reading. *Dallas News.* Retrieved from https://www.dallasnews.com/news/local-politics/2013/02/21/texas-students-outpace-rest-of-u.s.-in-math-science-lag-in-reading

Tomlinson, C. A. (2014). *The differentiated classroom: Responding to the needs of all learners* (2nd ed.). Alexandria, VA: ASCD.

Vygotsky, L. S. (1978). *Mind in society: The development of higher psychological processes.* Cambridge, MA: Harvard University Press.

Wilder, S. (2014). Effects of parental involvement on academic achievement: A meta-synthesis. *Educational Review, 66*(3), 377–397.

YouCubed at Stanford University. (2017). Tasks. Retrieved from https://www.youcubed.org/tasks/

Index

The letter *f* following a page number denotes a figure.

action research, essential components of, 151–154
algebraic manipulation, 33, 34–34*f*
assessment, expectations in, 123–124
assessment criteria, development of, 21
audit tasks
- data focused on, 146–147
- early and later work compared, 49*f*
- grades 3–5, 47–48
- grades 6–8, 48
- grades 9–12, 48
- instructional improvement and, 41
- primary grades, 47

base ten blocks, 22–23, 25*f*
bulletin boards, interactive, 132–133, 132*f*, 133*f*, 134*f*, 135, 136*f*

change, monitoring
- indicators for, 140*f*
- interactive bulletin boards for, 132–133, 132*f*, 133*f*, 134*f*, 135, 136*f*
- suggestions for, 136–139

change, supports for effective, 2
change, sustaining
- data focused on principals, 147–151
- data focused on students, 141
- data focused on teachers, 142–143, 146–147
- data's role in, 141

classroom management, 87–91, 88*f*
collaboration
- look-fors, 155
- student, 68–69, 155
- teacher, 128–129, 157

Common Core State Standards
- for Mathematical Practice, 6–7
- progressions, 81

common task samples, grades 1–5
- analysis criteria, 23*f*
- base ten blocks, 22–23, 25*f*
- decimal grids, 24, 26*f*
- fraction strips and grids, 24, 27–28*f*
- number line, 22, 24*f*
- ten frames, 23, 26*f*

common task samples, grades 5–8
analysis criteria, 29*f*
double number line, 28, 30*f*
graphs, 29, 31*f*
hundredths grid, 29, 30*f*
number line, 28
tape diagrams, 29, 31*f*
common task samples, grades 9–12
algebraic manipulation, 33, 34–34*f*
analysis criteria, 32*f*
graphs, 33, 34*f*
tables of values, 33
confidence
by principals, in teachers, 159
by teachers, in students, 99–102, 101*f*, 158
principals, 157
student, 82–86, 156
creativity, student, 85–86
curriculum, knowledge of and attention to, 112–118

data
anecdotal, basing decisions on, 19
change, role in sustaining, 141
focused on teachers, 146–147
principal focused, 147–151
qualitative, examples of, 18
quantitative, examples of, 17
relevant, collecting, 16–17, 19
schoolwide, collecting for change, 20–21
student focused, 141
teacher focused, 142–143
data, common task samples
grades 1–5, 22–28
grades 5–8, 28–31
grades 9–12, 32–35
data analysis, guiding questions, 35–36
decimal grids, 24, 26*f*
diagrams, 21
double number line, 28, 30*f*

engagement, 156
engagement, student, 77–82
enthusiasm, student, 78–79
expectations
in assessment, 123–124
high, principals exhibiting, 160
high, teachers exhibiting, 160
in instruction, 122–123

feedback, 108–109
fraction strips and grids, 24, 27–28*f*

graphs, 29, 31*f*, 33, 34*f*

homework review, 89
hundredths grid, 29, 30*f*

independence, student, 83
instruction
differentiating, 126–127
expectations in, 122–123
learning time, thoughtful use of, 87–91, 88*f*, 158, 159
instructional needs, attending to
student, 124–127, 160
teachers, 161

learning supports, 102–110, 158, 159
learning time, thoughtful use of, 87–91, 88*f*, 158, 159
learning tools, 72–77, 72*f*, 155, 157
lesson styles, 91–97

Making Math Meaningful to Canadian Students, K-8 (Small), 81
manipulatives, using, 72*f*, 74–77, 75*f*, 76*f*

math
  highlighting the important, 109–110
  making sense of, 82
  schools with great, 7, 8*f*, 9, 10–11*f*
math challenges, 60–62
math culture, conversations outside the classroom to build a strong
  assessment, expectations in, 123–124
  collaboration, 128–129
  curriculum, deep understanding of, 112–113
  curriculum, knowledge of and attention to, 112–118
  instruction, differentiating, 126–127
  instruction, expectations in, 122–123
  professionalism, 127–130
  professional learning, 128
  reflection on practice, 118–121
  resource choices, 114–115
  standards, focus on practice as well as content, 116–118
  structure, differentiating, 125
  students' instructional needs, attending to, 124–127
  taking responsibility, 129–130
  trying new strategies, 128
math culture, observing teachers to build a strong
  confidence in students, 99–102, 101*f*
  feedback, 108–109
  fostering deeper understanding, 97–99
  highlighting the important math, 109–110
  learning supports for students, 102–110
  lesson styles, 91–97
  listening to students, 102–104
  posing problems that pique curiosity or lead to generalizations, 96*f*
  probing not telling, 104–107
  problems, selecting to pique curiosity or to lead to generalizations, 94–96
  providing extensions for strong students, 107–108
  scaffolding not telling, 104–107
  teaching through problem solving, 92–94
  thoughtful use of learning time, 87–91, 88*f*
math culture, students in building a strong
  collaboration, 68–69
  confidence, 82–86
  creativity, 85–86
  engagement with math, 77–82
  enthusiasm, 78–79
  independence, 83
  interactions, 64–71
  learning tools, 72–77, 72*f*
  making sense of math, 82
  math talk, 65–67
  perseverance, 83–85
  types of questions asked, 69–71
  using manipulatives, 72*f*, 74–77, 75*f*, 76*f*
  using technology, 73, 73*f*, 74*f*
math culture, teachers in building a strong
  building support systems, 55–56
  parent buy-in for, 59–63
  resistance, overcoming, 57–58, 57*f*
  teacher buy-in, 56–58
math instruction
  belief survey, 15, 16*f*
  current practices in, 15
math talk, 65–67
Meet the Teacher Night, 60
models, 21

National Governors Association, 1
NCTM (National Council of Teachers of Mathematics), 5–6
number line, 22, 24*f*, 28

parents, involving, 59–63
perseverance, student, 83–85

principals
action research, using, 151
building collective consensus, assessing, 147–148
clarity of vision, articulating, 148–149
confidence, 157
confidence in teachers, 159
data focused on, 147–151
Principles and Standards for School Mathematics (NCTM), 5
Principles to Actions (NCTM), 2
probing, 104–107
problems
posing good, 89–90
selecting to pique curiosity or to lead to generalizations, 94–96, 96*f*
problem solving, 92–94
professionalism, 127–130, 160
professional learning
analysis criteria example, 46*f*
conversations outside the classroom, 128
focus of, 38
making decisions about, 36–38
motivators for, 41, 42–45*f*
summary charts, 38, 39–40*f*
targeted, 38, 40
professional learning, audit tasks
early and later work compared, 49*f*
grades 3–5, 47–48
grades 6–8, 48
grades 9–12, 48
instructional improvement and, 41
primary grades, 47
The Progression of Addition and Subtraction (Fletcher), 81

questions
asked by students, 155
asked by teachers, 157
to best meet standards, 12–13, 13*f*
doing and understanding, 99*f*
open, 95, 96*f*, 100
probing, 104–105
recall vs. understanding, 97–98*f*

reflection on practice, 118–121, 160–161
resources
to best meet standards, 14–15
conversations on choices of, 114–115
responsibility, teachers taking, 129–130

scaffolding not telling, 104–107
social media, 60, 61*f*
stakeholders, showing student work to, 138
standards
focus on practice as well as content, 116–118
meeting the spirit of, 9, 11
questions to best meet, 12–13, 13*f*
resources to best meet, 14–15
teachers' understanding of, 112–113
strategies, trying new, 128
structure, differentiating, 125
students
data focused on, 141
instructional needs, attending to, 124–127
listening to, 102–104
strong, providing extensions for, 107–108
teacher confidence in, 99–102, 101*f*, 158
students, building a strong math culture with
collaboration, 68–69
confidence, 82–86
creativity, 85–86
engagement with math, 77–82
enthusiasm, 78–79
independence, 83

interactions, 64–71
learning tools, 72–77, 72*f*
making sense of math, 82
math talk, 65–67
perseverance, 83–85
types of questions asked, 69–71
using manipulatives, 72*f*, 74–77, 75*f*, 76*f*
using technology, 73, 73*f*, 74*f*
summary charts, 38, 39–40*f*
summer numeracy challenges, 60–62

tables of values, 33
tape diagrams, 29, 31*f*
teacher conversations outside the classroom to build a strong math culture
assessment, expectations in, 123–124
collaboration, 128–129
curriculum, deep understanding of, 112–113
curriculum, knowledge of and attention to, 112–118
instruction, differentiating, 126–127
instruction, expectations in, 122–123
professionalism, 127–130
professional learning, 128
reflection on practice, 118–121
resource choices, 114–115
standards, focus on practice as well as content, 116–118
structure, differentiating, 125
students' instructional needs, attending to, 124–127
taking responsibility, 129–130
trying new strategies, 128
teachers
data focused on, 142–143, 146–147
instructional needs, attending to, 124–127
making performance predictions, 142–143
principals' confidence in, 159
teachers, building a strong math culture with
buy-in, getting, 56–58
parent buy-in for, 59–63
resistance, overcoming, 57–58, 57*f*
support systems, 55–56
teachers, observing to build a strong math culture
confidence in students, 99–102, 101*f*
feedback, 108–109
fostering deeper understanding, 97–99
highlighting the important math, 109–110
learning supports for students, 102–110
lesson styles, 91–97
listening to students, 102–104
posing problems that pique curiosity or lead to generalizations, 96*f*
probing not telling, 104–107
problems, selecting to pique curiosity or to lead to generalizations, 94–96
providing extensions for strong students, 107–108
scaffolding not telling, 104–107
teaching through problem solving, 92–94
thoughtful use of learning time, 87–91, 88*f*
Teaching Profession Act, 127
technology, using, 73, 73*f*, 74*f*
ten frames, 23, 26*f*
three-act math, 78
trust, five tiers of, 57, 57*f*

Which One Doesn't Belong? 78–79

About the Authors

Marian Small, the former dean of education at the University of New Brunswick in Canada, writes and speaks about K–12 math around the world. Her focus is on teacher questioning to get at the important math, include and extend all students, and focus on critical thinking and creativity.

Small has been on the author team of seven text series at the K–12 texts. Some professional resources she has written include *Making Math Meaningful to Canadian Students, K–8; Big Ideas from Dr. Small* (at several levels); *Good Questions; More Good Questions; Eyes on Math; Gap Closing* (for the Ministry of Education in Ontario); *Leaps and Bounds Toward Math Understanding* (at several levels); *Uncomplicating Fractions to Meet Common Core Standards in Math, K–7; Uncomplicating Algebra to Meet Common Core Standards in Math, K–8; Building Proportional Reasoning Across Grades and Math Strands, K–8; Open Questions for the Three-Part Lesson* (at several levels); and *Teaching Mathematical Thinking.* She is currently authoring *MathUp*, a new digital teaching resource, and *Fun and Fundamental Math for Young Children* for preschool and primary teachers.

Doug Duff is currently a principal instructional leadership coach with the Thames Valley District School Board in London, Ontario, Canada. He has been particularly interested in math education and school improvement for 20 years in various capacities, both nationally and internationally. He has worked with a major Canadian publisher as a senior consultant, multigrade textbook author, and national presenter for mathematics leadership and professional learning.

Duff has authored a web-based program, *MathUP School,* to support principals in building a math culture in their schools with Rubicon Publishing.

Related ASCD Resources

At the time of publication, the following resources were available (ASCD stock numbers in parentheses).

Print Products

Building a Math-Positive Culture: How to Support Great Math Teaching in Your School (ASCD Arias) by Cathy L. Seeley (#SF116068)

Making Sense of Math: How to Help Every Student Become a Mathematical Thinker and Problem Solver (ASCD Arias) by Cathy L. Seeley (#SF116067)

Teaching Students to Communicate Mathematically by Laney Sammons (#118005)

Unpacking Fractions: Classroom-Tested Strategies to Build Students' Mathematical Understanding by Monica Neagoy (#115071)

For up-to-date information about ASCD resources, go to www.ascd.org. You can search the complete archives of *Educational Leadership* at www.ascd.org/el.

ASCD myTeachSource®

Download resources from a professional learning platform with hundreds of research-based best practices and tools for your classroom at http://myteachsource.ascd.org/

For more information, send an e-mail to member@ascd.org; call 1-800-933-2723 or 703-578-9600; send a fax to 703-575-5400; or write to Information Services, ASCD, 1703 N. Beauregard St., Alexandria, VA 22311-1714 USA.